Fast Reactors

A Solution to Fight Against Global Warming

Joel Guidez

Academic Press is an imprint of Elsevier
125 London Wall, London EC2Y 5AS, United Kingdom
525 B Street, Suite 1650, San Diego, CA 92101, United States
50 Hampshire Street, 5th Floor, Cambridge, MA 02139, United States
The Boulevard, Langford Lane, Kidlington, Oxford OX5 1GB, United Kingdom

ISBN 978-0-12-821946-1

For information on all Academic Press publications
visit our website at https://www.elsevier.com/books-and-journals

Publisher: Charlotte Cockle
Editorial Project Manager: Zsereena Rose Mampusti
Production Project Manager: Sreejith Viswanathan
Cover Designer: Vicky Pearson Esser

Typeset by STRAIVE, India

Contents

About the author

Joel Guidez began his career in the field of sodium-cooled fast reactors after graduating from the École Centrale de Paris in 1973. He worked at Cadarache for 8 years on the design, dimensioning, and testing of sodium components for Superphénix, followed by the initial results, in his field, from the Phénix sodium-cooled fast reactor that started up in 1974. He then joined Phénix, where for 5 years he would be in charge of measurements and tests on the power plant.

In 1987, he returned to Cadarache to head a thermohydraulics laboratory, where many tests would be performed for Phénix, Superphénix, and the European Fast Reactor (EFR) project. After a period of apparent unfaithfulness to fast reactors, during which he successfully managed the OSIRIS research reactor located in Saclay and the European Commission's reactor, HFR located in the Netherlands, he returned to Phénix in 2002, where he would manage the reactor until 2008 during its final operating phase.

His first book "Phenix. The Experience Feedback" was published in 2012, translated into English in 2013, and republished in French the same year. His second book "Superphenix. Technical and Scientific Achievements" was published in 2015, translated and published in English in 2016, and republished in French in 2017.

He was in charge of the design of the European ESFR SMART sodium fast reactor project from 2017 to 2021 and began to work actively on molten salt reactors in 2018 by organizing a 2-day seminar on the subject with all the French specialists.

Since 2020, Mr. Guidez has retired from CEA and works as an independent expert in his company *Nuclei cogitum.*

Foreword

Global warming is a concrete reality that is experienced every day around the world and whose effects will accelerate in the years to come. The main cause, responsible for 70% of this phenomenon, is known: the immoderate use of fossil fuels, viz. the "pickaxe" culture that has led us to quench our infinite thirst for growth using energy resources from the Earth (coal, oil, and gas). The solution has also been known for years: produce less CO_2, and in particular stop using these fossil fuels, which are the biggest emitters.

In this context, nuclear power, the energy source that produces the least amount of CO_2, clearly has an important role to play. Unfortunately, current nuclear power technology suffers from serious disadvantages that prevent it from fully playing this essential role in the future of our planet. The first of these is the huge size of current reactors, which results in construction times of at least around 10 years and the requirement for very significant investments. Meanwhile, their operational complexity also limits their applicability in many countries.

Therefore, new, smaller and smarter technologies deserve a closer look, including small modular nuclear reactors such as the SMR (Small Modular Reactor), AMR (Advanced Modular Reactor), MMR (Micro Modular Reactor), or XSMR (extra Small Modular Reactor). Whatever the power generation capacity of such installations, their common feature is the industrialization of their production, the modularity intrinsic to their design, and a decentralized vision of electricity production, enabling in particular the supply of electricity to isolated regions or where the power grid infrastructure is of poor quality [1].

In addition, many countries reject nuclear power, mainly for two reasons: the risk of accidents, and the creation of very long-lived waste. However, these new types of reactors, known as fourth-generation technology, will make it possible to envisage models that could address these problems, that is, fast reactors.

Joel Guidez has written two remarkable books on reactors using this technology: one on feedback from the Phenix sodium fast reactor, and one on the achievements of the Superphenix sodium fast reactor. These books have been distributed in French and English versions and published in new editions when supplies ran out. Mr. Guidez is one of the best-known international experts in this field.

In this new book, Mr. Guidez recalls the fundamental advantages of fast reactors that make it possible to operate them using available fuels and in particular to burn long-lived waste, in order to ultimately produce only short-lived waste that is perfectly manageable. This will avoid the need for uranium mines and enrichment plants, and enable operation for thousands of years with existing stocks of these products that are already stored and available. Note that, without the hope offered by these fast reactors, such products would become long-lived nuclear waste.

One may ask why such reactors, which have been built in France (Phenix and Superphenix) as well as Germany, the United Kingdom, the United States, Russia, China, India, and Japan, have not invaded the world? An entire chapter of this book explains the reasons for this, providing the means to analyze and overcome these difficulties.

As President of the Confrontations Europe think tank and the House of Europe in Paris, I was particularly interested in the chapter on the latest studies on the European ESFR SMART project, which, based on feedback from sodium reactors, reanalyzed their design to further improve their safety. This analysis is remarkable because this improvement was achieved not by adding systems and making the reactor more complex but on the contrary by simplifying it. This leads to a reactor with very simple operation, without pressure, whose safety is ensured passively through natural convection. For example, the removal of the residual power (the cause of the explosion at Fukushima) is achieved here via natural convection, by simply opening hatches.

Note that all these simplifications in the concept are even more readily applicable to small reactors. This drive toward the SMR is already the subject of a new 4-year European program that will begin this year and should lead to versions of the SMR that could be manufactured in a factory.

However, Joel Guidez does not stop at sodium fast reactors; this book also analyzes all the possible versions of the fast reactors proposed by the Generation IV International Forum (GIF). It is concluded that fast molten salt reactors are another very interesting possibility for the future. The organization in 2018 of a 2-day seminar in Massy with all the French actors working on this subject enabled him to confirm the very significant potential advantages of such reactors in terms of safety, the ease of the fuel cycle, and waste management. In particular, regarding safety, the avoidance of the possibility of a serious accident could facilitate the social acceptance of such reactors

Unfortunately, feedback regarding experience with this type of reactor is sparse and they therefore require significant developments. These are currently underway, in particular through the French start-up NAAREA, which offers a particularly innovative double technological leap by combining SMR and Generation IV with a fast molten salt reactor.

There are therefore two technological possibilities currently: sodium reactors using proven technology, which we know how to build with all the simplifications/improvements in terms of safety, and molten salt reactors, which offer significant potential advantages (in terms of safety, cost of mass production, a simpler fuel cycle, etc.) but that lack feedback and require development.

Fast-reactor technology is therefore a real hope for humanity. It would enable the production of almost unlimited amounts of energy by burning long-lived waste but without emitting CO_2 or requiring uranium mines, and with a remarkable level of safety.

Let us conclude like the author: Will humanity be able to take advantage of this opportunity? The climate emergency is here [2], and the major technological choices cannot wait!

Michel Derdevet
President of Confrontations Europe
President of the House of Europe in Paris
Professor at the Institute of Political Studies in Paris
and at the College of Europe in Bruges

References

[1] Cf. Revue de l'Energie n°657, juillet-août, 2021, pp. 52–64.
[2] Dans l'urgence climatique. Penser la transition énergétique, Gallimard, 2022. mars.

Introduction

For the serious reader, there are today two categories of books: those that focus on the content and aim to communicate a particular knowledge, and those that focus on the structures and modes in which facts are organized.

The first two books I wrote on the subject of fast sodium-cooled reactors clearly fell into the former category. Their titles "Phénix Experience Feedback" and "Superphénix Technical and Scientific Achievements" clearly show a desire to capitalize on knowledge in this sector of fast sodium-cooled reactors and to transfer this knowledge to the reader.

This third book falls into the latter category. It also informs on this sector of fast reactors, but less to communicate knowledge than to try to understand why this technology, which has the potential to solve the energy problems of the planet while addressing the ecological challenges that await humanity in this field, is currently in a situation of near failure, with very low development.

A historical assessment of these reactors indeed shows a situation of abandonments, shutdowns, and postponed projects. An analysis of the reasons for this failure is then carried out, reasons that are linked largely to modes of societal organization and to structural reasons specific to each country. Although this analysis reveals obstacles, some difficult to remove, the book continues with a review of the current state of knowledge on these fast reactors and offers some proposals for the future.

One chapter outlines the status of sodium-cooled fast reactor development in Europe with the ESFR-SMART project.

In addition, developments are proposed on molten salt reactors, which offer very interesting new potential for the future of these fast reactors.

We were a small group of French engineers who, in the 1970s, understood the potential of a technology capable of producing energy for thousands of years with waste already available and stored—energy that does not produce CO_2, does not produce dust, does not need a uranium mine, and works with waste already available and transforms it into waste that has a reduced duration of radiotoxicity and is therefore manageable.

We are sorry for the successive abandonments, the messes, and the extent of media lies on this subject. I hope this book will restore some truth and hope in this sustainable clean energy.

Dream or reality? The reader will remain the sole judge.

Joel Guidez

CHAPTER

Where do *Homo sapiens* get their energy?

1

Abstract

After outlining how humanity's energy needs have evolved historically, we explain the main consequences of the recent and massive use of fossil energies, that is, global warming, linked to CO_2 emissions.

Some short-term consequences of the current global warming are summarized, including the rise in sea levels, the disappearance of the poles, climate change, desertification, the acidification of the oceans, the increase in extreme climatic events, etc.

There is therefore an urgent need to reduce the use of fossil fuels. However, these currently provide around 85% of the world's energy. Moreover, prospective studies show that, despite efforts to save energy, energy consumption will continue to grow owing to several factors: population increase, catch-up of populations in poor countries, and new needs related to digital technology or new technologies (e.g., electric cars, hydrogen).

The only current and operational alternatives that do not produce CO_2 and can replace these fossil fuels are nuclear power and renewable energies: mainly solar, wind, and hydroelectric. One can mention many current research projects investigating the use of fusion, geothermal energy, CO_2 capture, the tides, etc. However, these energies are not operational today and perhaps never will be, and therefore will not be useful for the development of the necessary short-term response.

The limits of renewable energies are outlined. In particular, solar and wind power cannot be controlled. As we do not know how to store large quantities of electricity, they cannot alone meet the electrical needs of a network. In addition, they are diffuse energies, requiring large surfaces, which poses other ecological problems. The example of Germany is given, which experienced difficulties with its nuclear exit policy despite massive use of renewable energies.

Only nuclear power can provide the essential and carbon-free complement to these renewable energies.

But the energy needs are immense and far exceed, even granting a large share to renewables, the possibilities of current nuclear power. In addition, certain problems must then be solved, such as the supply of uranium and the management of waste.

Fast Reactors. https://doi.org/10.1016/B978-0-12-821946-1.00004-8

It is therefore necessary to analyze the possibilities of new types of reactors, capable of solving these problems and ensuring long-term clean, carbon-free, and quasirenewable energy for *Homo sapiens*.

Keywords

Energy, Global warming, Fossil energy, Renewable, Nuclear

Some history

Earth was formed about 4.5 billion years ago. It was then certainly a radioactive broth, with all the constituents of Mendeleev's periodic table. However, the half-life of the radioactive elements being what it is, even the long-lived elements with a half-life on the order of 80 million years, like plutonium-244, have almost disappeared. (It is estimated that in a flowerpot with a kilo of soil there are about 10 becquerel of plutonium left.) Uranium would also have long since disappeared if two of its isotopes, uranium-238 and uranium-235, had not survived on Earth to the present day owing to their exceptionally long lifespans. The half-life of uranium-238 is 4.5 billion years. That of uranium-235, which disappears faster, is "only" 700 million years. While the abundance of the two isotopes was initially similar, natural uranium today consists of 99.3% uranium-238, compared with 0.70% uranium-235. The nuclei of uranium-235 and uranium-238 are, with those of thorium-232, the heaviest existing in nature. They were formed a very long time ago, during the explosion of very large stars called supernovae. Uranium-235 therefore remains the only fissile isotope available in nature to cross paths with *Homo sapiens*, who arrived much later, and today this isotope enables us to produce nuclear energy, the subject of this book.

Homo sapiens indeed appeared a very short time ago, around 150,000 years. Within a few tens of thousands of years, they succeeded in pushing the other human species to extinction and then colonized the entire planet from Africa to Europe, from Bering Strait to Tierra del Fuego, and from Asia to Australia. As a source of energy, besides the heat of the sun and the force of the wind in their sails, they could count only on their muscular strength and on the combustion of wood. This source of energy remains today a very important resource for many populations.

Sedentarization by agricultural transition was initiated in Mesopotamia around 9000 years before our era. It spread, creating larger and larger agglomerations, more and more complex societies, and new energy needs. The use of animal energy and slavery spread to provide additional energy that became necessary, at an economically affordable rate.

The arrival in the 19th century of the industrial revolution allowed, through the use of fossil fuels (coal, oil, gas, etc.), an explosion of the energy available to humans. It was this availability that led to a spectacular development of society and, along the way, to the abolition of slavery, which had become less and less useful.

FIG. 1

View of an industrial landscape.

Nuclear power and renewable energies were introduced after the Second World War, but currently more than 85% of humanity's energy is still produced by these fossil fuels, that is, by the combustion of stocks of carbonaceous products accumulated over millions of years: coal, oil, gas, lignite, etc. New techniques are being developed to access fossil fuels, such as the extraction of shale gas by hydraulic fracturing or the capture of methane nodules at the bottom of the sea.

Oil and gas pipelines continue to be pulled around the world as the planet burns (Fig. 1).

Global warming

This intensive use of fossil fuels leads to the release of colossal quantities of CO_2 emitted during their combustion. Since 1850, this release of CO_2 has led to an exponential increase of the CO_2 content in the air across the globe. This presence of CO_2 induces a greenhouse effect already widely described in the 19th century, leading to global warming of the planet.

The Intergovernmental Panel on Climate Change (IPCC) was created to analyze these effects. Since 1990, it has periodically produced successive reports that show a predictable and almost inevitable evolution of this global warming.

Without reproducing here all the conclusions of the last report of 2021 [1], to appreciate the stakes, it is necessary to outline the main consequences of a 4°C rise in temperature, a scenario expected to occur if the consumption of fossil fuels continues at the current rate.

- Melting of glaciers and poles
  In 30 years, the Arctic sea ice will have shrunk by 2 million km^2. Projections show that it could disappear by 2050. In addition, glaciers are retreating and a large number will have disappeared by this date (Fig. 2).
- Rising waters
  The rising waters are due to two factors: the expansion of the warmer sea and the melting of the ice of Greenland and Antarctica. Projections show that by 2050 entire areas, now highly populated, will have disappeared under water. In 2100, sea level rises of several meters will upset all coastal areas (Fig. 3).
- Desertification and climate change
  Whole regions in Africa, but also on other continents, will become deserts. Sea currents will change with the warming of the poles, and the climates of the regions concerned will be modified. Millions of people will be forced to migrate.
- Exceptional climatic events
  The warming of surface waters will lead to an increase in exceptional rainfall, violent floods, cyclones, etc., an increase that is already perceptible today. In addition, intense heat waves above 50°C are occurring more and more frequently, often accompanied by corresponding forest fires. These fires are increasingly violent and difficult to control (Fig. 4).

FIG. 2

Iceberg resulting from the melting of the North Pole.

FIG. 3

Illustration of rising water.

FIG. 4

Illustration of increasingly intense forest fires.

- Ocean acidification
  The ocean absorbs about half of the CO_2 emitted, leading to its acidification. Its pH of 8.15 in 1950 has fallen to 8 today. It will fall to 7.6 in 2100 if CO_2 emissions continue as usual. Already, coral reefs are seriously threatened, as well as many species of plankton, the basis of the food chain in the oceans (Fig. 5).

FIG. 5

The Great Barrier Reef in 2002 and 2014.

Worse threshold effects exist, such as the disappearance of ice from the poles announced in the short term, the release of the methane contained in the permafrost when it thaws, and the disappearance of certain marine currents. These threshold effects could lead to runaway disruption with characteristics of nonreversibility. How do we re-create the North Pole, the storage of methane in permafrost, or a disappeared Gulf Stream?

All these consequences are announced for the short term, that is, a few decades, a timespan that directly concerns our children. And they are already starting to be very visible (Fig. 6).

FIG. 6

Young people and global warming.

The planet will survive these changes without any problem, but the life of *Homo sapiens* risks being totally disrupted.

What are the energy needs of the future?

Notwithstanding the slogan "The best energy is that which we do not consume," it seems very difficult to reduce our energy consumption. Strategies such as better insulation of homes or better class appliances bring slight improvements that should not be overlooked but have little overall effect.

Two examples show the difficulty of reducing a country's electricity consumption.

After the Fukushima nuclear accident, Japan shut down all its nuclear power plants, which had produced about 30% of the country's electrical energy. Being an island and unable to connect to neighbors, measures were taken to reduce consumption: elimination of public lighting and moving walkways in the subway, call for good citizenship, etc. However, all this led to a temporary reduction in energy consumption of only about 5%.

The other example is Germany, which decided to start shutting down its nuclear power plants, with an ambitious long-term energy plan based in part on a desire to reduce energy consumption by around 25% compared with 2008. Twelve years later, despite significant and continuous efforts, consumption has only stabilized (Fig. 1) The 4% drop between 2019 and 2020 is cyclical and mainly due to the coronavirus health crisis.

The assumption of the 2010 energy concept, which still counted on a 25% decline in gross electricity consumption by 2050 compared with 2008 (~614 TWh) was modified in 2020/21. Following the increased electrification of other sectors of the economy, the German Government now expects a sharp increase in electricity consumption. The new government in office since December 2021 expects gross consumption of 750 TWh in 2030. According to various studies, electricity consumption will exceed 1000 TWh in 2045 with a peak consumption of more than 100 GW (Fig. 7).

Prospective studies show that world energy consumption will inevitably increase [2–6]. The scenarios can be very different, but all predict a significant increase in energy demand.

This increase in demand is due to three main reasons: the demography, the energy catch-up desired by poor countries, and the new needs.

Regarding demography, the world population was 2.5 billion in 1950, exceeds 7 billion today, and is expected to exceed 10 billion in the short term. Each day there are 228,000 more humans on the planet (Fig. 8).

In terms of needs, billions of people have insufficient access to energy and want to reach at least the minimum level necessary for their development.

Moreover, energy consumption linked to new digital practices is increasing almost exponentially. In addition, some new technologies such as electric cars or hydrogen technologies are also energy intensive.

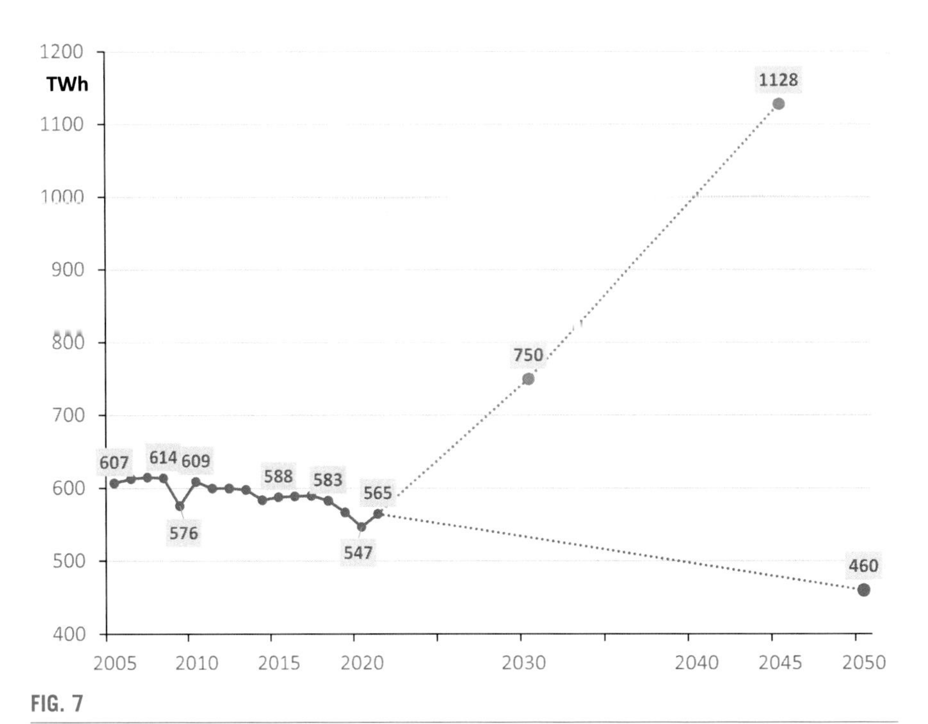

FIG. 7

Evolution of electricity consumption in Germany. In *blue*, the initial target. In *red*, the today target

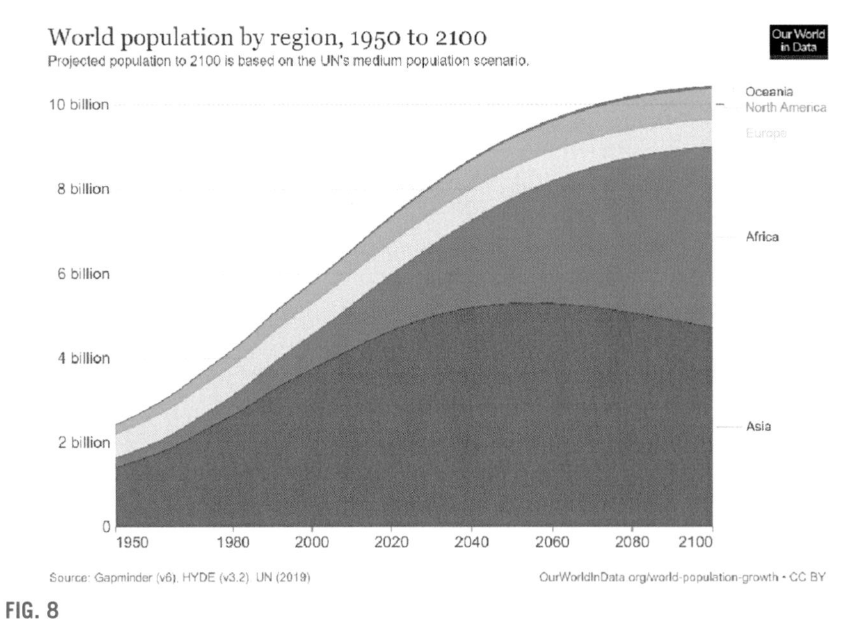

FIG. 8

Evolution of the world population.

The return of the electricity fairy?

Much of the fossil fuel is used for heating homes and for transportation.

For the heating of houses, better insulation and the use of passive solar technology allow the bill to be minimized. However, the final necessary complement is electric.

For transport, cars, buses, and trucks powered by electricity or even hydrogen are trying to take over. However, it takes 52 kWh to produce one kilo of green hydrogen. It is therefore still the electricity fairy that will be in great demand to replace fossil fuels in these applications. Nevertheless, the quantities of electricity needed to replace fossil fuels with electricity or hydrogen would be very large. For example, the replacement in France, of all the consumption of gasoline for cars, by a consumption of hydrogen produced by electrolysis, would require the doubling of the production of electricity.

It is still necessary that the fairy is not a witch and that this electricity is generated without producing CO_2!

Electrical energy without CO_2 production?

Currently, the electricity fairy is unfortunately an ugly witch, and most of the electricity in the world is produced from fossil fuels. For example, the 51 Chinese nuclear reactors account for only 5% of Chinese electricity production. In Africa, where the network is often deficient, a large part of the production is ensured by individual fuel-powered generators.

In fact, no electricity production technique avoids the production of CO_2, but the quantities vary significantly depending on the techniques used.

For fossils, the worst is coal and lignite, followed by fuel oil, then gas.

Nuclear and renewable energies do not produce CO_2 when they operate (direct emissions), but they produce it for their construction, dismantling, and supply (indirect emissions).

All this has been calculated and identified in numerous papers. The results may vary slightly depending on the method of calculation. For example, a solar panel made in China with coal-based energy will have a much higher carbon weight than a panel made in France with a mix of nuclear and renewable energy.

However, the main results are as follows (Fig. 9):

The transition to nuclear and renewable-based electricity production would allow extremely significant reductions in the production of CO_2.

The witch could therefore become a fairy again.

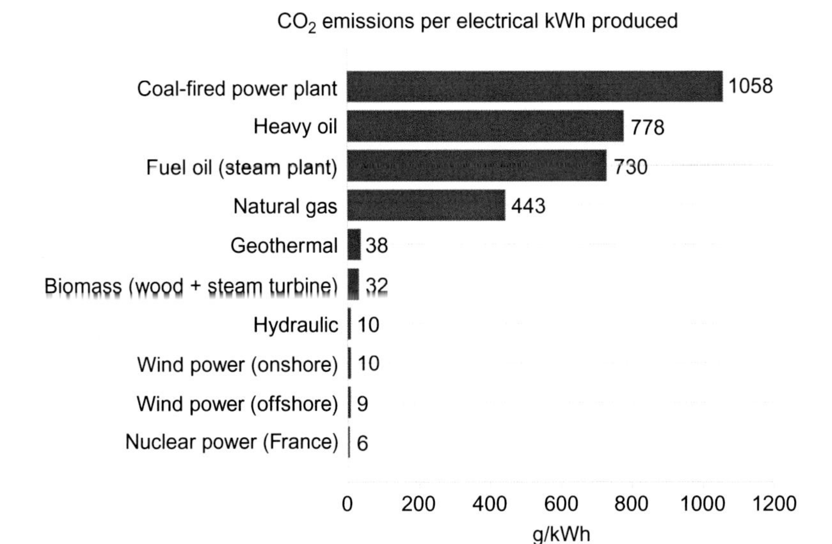

FIG. 9

CO_2 emissions in g/kWh for the different types of energy.

A world powered by renewables alone?

Some ecologists dream of a world powered by renewables alone, i.e., today mainly solar, wind, and hydropower.

This vision faces significant problems.

Cost is an issue, but could be overcome with political effort.

One of the main problems is the occupation of space. These energies are quite diffuse and require large spaces. If we place an EPR of 1600 MWe on approximately 1 km^2, it will be necessary to produce the same quantity of electricity at the end of the year and, taking into account the difference in performance, approximately 3000 wind turbines of 2 MW. If we count 0.1 km^2 per wind turbine, because of the necessary distance between them, we arrive very roughly at 300 km^2 covered with wind turbines about 200 m high. We immediately see the ecological consequences in terms of the destruction of landscapes, noise, massacre of birds, etc. for these important surfaces. For solar, with 160 kWh/m^2 per year, the same calculation gives a necessary area of about 70 km^2 of panel area. Germany, which has deployed a large number of these systems, already announced a coverage of 2% of its territory. However, this figure takes into account only the surface on the ground and not the surfaces affected by visual or auditory pollution.

Another problem is the random aspect of this production when we do not know how to store the current in large quantities and at an affordable price. The German example is very interesting in this regard. Colossal efforts have been made year after year to increase the number of renewables, replacing nuclear. At the end of 2021, renewable energy contributed 138 GW and produced 43% of the energy during the first three quarters. This is to be compared with the 63 GW of the French nuclear fleet, which produces a little over 70% of French electricity.

The maximum current demands are in winter and in the evening, when returning from work. Solar technology is no longer operational, and wind is random. The 138 GW of installed renewable energies are then of little use. The maximum historical German consumption is around 82 GW. In practice, this obliges Germany at the end of 2021 to maintain a controllable fleet of 88 GW. This fleet is mainly made up of fossil fuels and essential to avoid power cuts in the evening.

We arrive at a total fleet of 226 GW compared with 63 GW of French nuclear power, which produces more than 70% of electricity.

Because of this poor efficiency of renewables linked to the poor load factor and random operation, CO_2 production remains very high in Germany and has decreased by only 26% from 2000 (1043 Mt CO_2) to 2021 (772 Mt CO_2) (Fig. 10).

Germany remains the largest CO_2 emitter in Europe, despite this colossal effort in favor of renewable energies.

Worse, the values go up in 2021. The shutdown of two nuclear power plants at the end of 2021, which will be followed in 2022 by the closure of the last power plants, will lead to the elimination of the production of carbon-free nuclear electricity at the end of 2022. Nuclear plants still accounted for 11% of electricity production

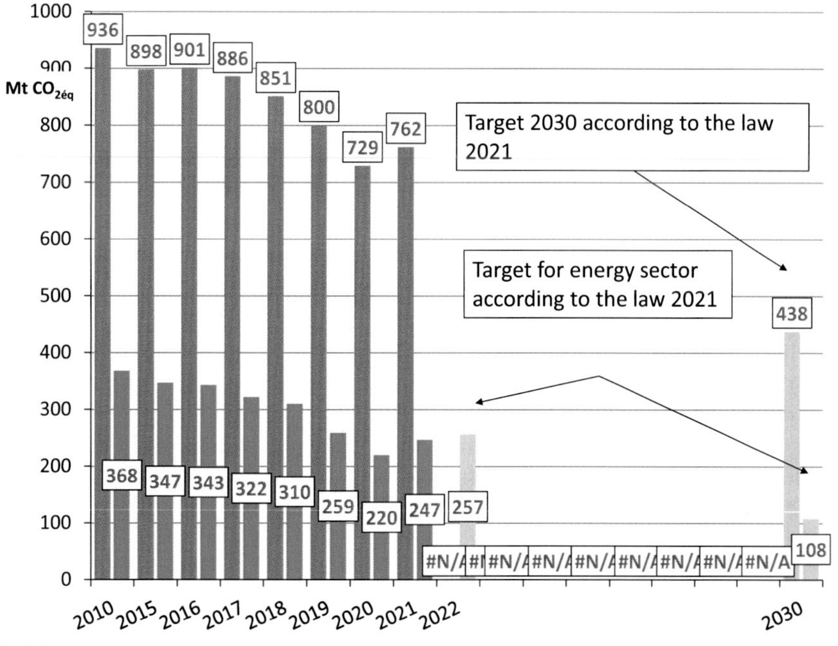

FIG. 10

Evolution of emissions in million tons of CO_2 equivalents per year from 2010 to 2021 in Germany.

in 2021 and cannot be replaced in 1 year by new renewable energies. A rise in CO_2 production therefore seems inevitable in 2023.

Continuing to increase renewables leads to a situation of overproduction when there is wind and sun at the same time. Episodes of power sales at negative prices are multiplying because of the significant installed renewable power that is dumping power when there is no demand.

Adding solar or wind power does not solve the problem of nonproduction on windless evenings. It therefore appears extremely unlikely to arrive at a situation where 100% of the current is produced by the renewable energies while avoiding periods of power cuts.

Wood is a special case in terms of renewable energy. It is the most catastrophic fossil fuel in terms of yield and therefore CO_2 production, as well as fine particle emissions. But it remains generally defended by ecologists, as a renewable resource. Indeed, if we replant the wood that we have just burned, after about 40 years the CO_2 balance will be zero. This reasoning assumes proper management of the forest, which is not the case in many places where deforestation is progressing. In these places, there is a double penalty: maximum production of CO_2 with removal of trees capable of CO_2 trapping.

New research for the future?

Many studies are, of course, underway on new forms of production without CO_2 production. We will cite examples and in a nonexhaustive way, given the proliferation of proposals:

- CO_2 capture
  In my opinion, it is often false advertising by fossil energy companies to make people believe that they will 1 day be able to capture and then inject the CO_2 they produce somewhere, and during this time continue their production. In any case, the capture alone increases the costs significantly. The reinjection of this CO_2 is tricky, and the final cycle is still poorly understood.
- Electric current storage
  Many studies and tests are taking place to find a means of storing current in thermal, chemical, and mechanical form. This remains very expensive, and CO_2 is produced somewhere in the process. The most interesting way seems to be the production of hydrogen by electrolysis, which would allow a mode of chemical storage while creating an alternative fuel, but at the cost of a significant drop in yield during the operation.
- Fusion
  The gigantic International Thermonuclear Experimental Reactor (ITER) project is at the plasma behavior demonstration stage. No kilowatts of energy will be produced there.
  If ITER is successful, the production would have to be studied in an even more colossal DEMO project with equally hypothetical results (Fig. 11).

FIG. 11

View of the ITER site in Cadarache (France).

- Geothermal
  Facilities are working, but there are no great possibilities in sight.
- Use of tides
  Sites for tidal power plants are rare. Submerged machines have prohibitive weight/production ratios, and their maintenance at sea is very complex.

In conclusion, after this extremely rapid and superficial scan, no strategy for the future can be established today on the basis of these studies, which are very interesting but remain at the level of prospective research. Unfortunately, those who do not want to act often use promises such as fusion or CO_2 capture to justify inaction.

We must continue this research, which will perhaps 1 day succeed, but we must also act quickly with operational solutions that work.

And nuclear?

Taking one of the scenarios for the growth of world energy demand by 2050, if we assume, in a utopian way, that the consumption of fossil energy stops, even if we multiply by 10 the renewable energies available, there remains a nuclear energy need that is increased 30–40-fold, or more, depending on the scenarios adopted.

It appears in this context that nuclear power is an essential basis for combatting global warming, in particular to ensure the production of electricity when renewable energy is lacking (at least every windless evening) and to ensure sufficient production for the new needs to come, replacing fossil fuels, particularly domestic heating and transport.

Currently, about 500 reactors, mainly pressurized water reactors, operate globally, producing about 10% of the world's electricity, without producing CO_2.

Nevertheless, certain factors have led to a lack of support from part of the population and the abandonment of nuclear power by certain countries. The main factors mentioned by antinuclear advocates are the risk of accident, waste management, and uranium supply.

Another important problem is the cost and schedule drift of these reactors. This drift is due to many factors, in particular a continuous increase in safety standards. The heaviness of the investments and the duration of the construction, which can exceed 10 years, are major obstacles to development. The prices of nuclear power, which were competitive with those of fossil fuels in the 1970s, have increased considerably today.

Let us recall two quotes from Bill Gates (November 2019):

> *The whole nuclear industry today is making a product that is too expensive and whose safety, while actually quite good, depends too much on human operators. So the nuclear industry will only survive if there is a new generation whose intrinsically safe design and economics are far better than anything currently available. Today's reactors are uneconomical. Ignore everything else. So the nuclear industry is going to die out, and that's true globally, unless there's a new design.*

If we add to the cost the complexity of current reactors, we see that many countries will not have access to them in the decades to come and will therefore continue to rely on fossil fuels.

The new nuclear?

However, there are new reactor designs that would theoretically make it possible, while retaining the advantages of combatting global warming, to respond to current nuclear problems. They do not require a uranium mine. They work with existing waste from conventional pressurized water reactors, and they produce more easily manageable short-lived waste.

It is no longer the electricity fairy but Merlin the enchanter!

Conclusion

Humanity requires energy, and the use of fossil fuels during the industrial revolution opened up immense possibilities. Unfortunately, the corresponding colossal releases of CO_2 will lead in the short term to intolerable upheavals for all of humanity and the planet.

All available projections show that energy-saving efforts, even if they are virtuous and desirable, will not reverse the trend and that the increase in global energy demand will continue in the decades to come.

It is therefore urgent to change our production methods and in particular to stop using fossil fuels. Their replacement by renewable energies can only be partial, mainly for reasons of cost, limited available space, and the current impossibility of storing these random energies. The example given by Germany is very valuable because it clearly shows, after more than 20 years of colossal efforts, the limits of these renewable energies.

Nuclear energy therefore clearly has a role to play in helping to replace these fossil fuels with a mode of production that can be controlled and that limits global warming.

Unfortunately, current nuclear power, based on pressurized water reactors, has several characteristics poorly perceived by the population, mainly in terms of waste management, accident risk and future uranium availability. Hence, it has been abandoned or limited in many countries. It also presents potential difficulties for the supply of uranium, in the long term.

New types of reactors would make it possible to respond to these technical problems while increasing their social acceptance. We outline these in the following chapters.

References

[1] GIEC, The Sixth Report of the GIEC/2021, 2021.
[2] https://www.iea.org/reports/electricity-market-report-december-2020/2020-global-overview-capacity-supply-and-emissions.
[3] http://www.realisticenergy.info/transformations/FCha/.
[4] https://www.planete-energies.com/fr/medias/chiffres/la-consommation-mondiale-d-energie-en-2019.
[5] https://www.enerdata.fr/publications/analyses-energetiques/bilan-mondial-energie.html.
[6] https://www.connaissancedesenergies.org/les-chiffres-cles-de-lenergie-dans-le-monde-170926.

CHAPTER

What are the advantages of fast reactors?

2

Abstract

As of 2021, approximately 500 nuclear power reactors are operating worldwide. Most of them are water reactors, that is, they use water as a coolant. Some reactors use a different cooling fluid that makes it possible to slow down neutrons less. These are called fast reactors.
This chapter explains the principle of these fast reactors with higher-energy neutrons, and all the related positive consequences for nuclear electricity production:

- No more need for uranium mining
- No more need for enrichment plants
- Several thousand years of energy production using only existing and already available nuclear waste such as depleted uranium issued from enrichment activities and plutonium issued from reprocessing activities
- Other possibilities using thorium
- Isogenerators, burning only depleted uranium
- Production without CO_2 emission, making it possible to limit global warming
- Production without dust emission and without chemical discharges during reactor operation
- Production using existing nuclear waste, reducing the quantity of nuclear waste
- Possibility to burn all actinides and to reduce drastically the duration of final nuclear waste radiotoxicity, making the final waste easier to manage because it becomes harmless in a few hundred years

Compared with renewable energies such as solar or wind, it appears that the ecological footprint of nuclear energy is smaller.

Keywords

Fast reactor, Breeder, Plutonium, Uranium, Thorium, Waste, Isogenerator

Fissile and fertile material

An isotope is said to be fissile if its nucleus can undergo nuclear fission under bombardment by neutrons of all energies (fast or slow). The only fissile available in nature is ^{235}U, which is found in natural uranium in a proportion of about 0.7%, the main other isotope being ^{238}U (99.3%).

Fast Reactors. https://doi.org/10.1016/B978-0-12-821946-1.00003-6

Fertile material is material that, by capturing neutrons, creates new fissile material. In this manner, fissile elements are created in the fuel during the operation of the reactor, by neutron capture. The most important by mass is the plutonium created by the capture of neutrons in ^{238}U. There are also other minor actinides created: neptunium, americium, and curium. Thorium is another fertile element, which, in neutron flux, will create the fissile element ^{233}U. All these new fissile elements will participate in the energy production in the reactor. Some of these products, if extracted by reprocessing, can also enrich the range of fissile products available.

In conclusion, the only fissile element available in nature is ^{235}U. Its use in a fission reactor, in conjunction with fertile elements, allows the creation of new fissile elements that are potentially usable, the most useful being plutonium.

Reminder on the principle of fast reactors

When a neutron is ejected during fission of a fissile, it is animated with great speed and great energy. It is called "fast." With this energy, it cannot crack the uranium-235, because the "fission cross-section" (measured in barn) of uranium-235 is very low for fast neutrons; in other words, the probability of producing a fission is very low. However, this probability increases with the slowing down of the neutrons, and the neutrons naturally slow down by the many shocks they undergo on all the other nuclei of the core. When they hit a big nucleus, they bounce back without losing much speed, but when they hit a light nucleus, such as hydrogen, they slow down very effectively.

For this reason, light nuclei are introduced into the core of many reactor concepts. These light nuclei are called "moderators," and include water in pressurized water reactors (PWRs) and carbon in the first French sector of gas-cooled natural uranium reactors or in the Chernobyl RBMKs, among many others (e.g., heavy water in Canadian CANDU reactors).

When the neutrons arrive at the slowest speed corresponding to the thermal agitation of the fuel cores, they are called "thermal" and the fission cross-section is then maximum. However, the capture cross-sections (absorption of a neutron without producing fission) also increase with the slowing down of the neutrons, which has other consequences.

During the operation of a reactor, where the power is produced initially by the fissions of ^{235}U, the neutron captures in the nonfissile ^{238}U create fissile plutonium (mainly ^{239}Pu), partially replacing the cores of ^{235}U, which have disappeared. As a result, the fuel in the reactor wears out much less quickly, and the energy extracted from it is therefore significantly higher than the energy that could have been extracted from ^{235}U alone. We can even recover ^{239}Pu from spent cores and turn it into a new fuel where ^{239}Pu replaces ^{235}U as the fissile element, which is called mixed oxide (MOX).

However, ^{239}Pu also has the property of capturing neutrons and transforming into ^{240}Pu, which is no longer fissile, and constitutes a neutron poison. The mixture of ^{239}Pu and ^{240}Pu very quickly becomes unusable in this type of thermal reactor.

In contrast, in fast neutron fluxes, sterile neutron absorption in plutonium is practically zero, and the ^{239}Pu generated by capture in ^{238}U remains pure enough to be cracked. As the fission cross-sections in fast fluxes are also lower, a significantly higher enrichment in fissile nuclei ^{239}Pu (and ^{235}U if applicable) is required. Still other constructive measures are necessary to save neutrons and prevent their neutrons from slowing down. In particular, coolants without hydrogen are needed, such as sodium, lead, and helium.

We then arrive at the concept of “fast” reactors (more exactly, “fast neutrons”) whose fuel is a mixture of ^{239}Pu and ^{238}U. This ^{238}U produces, by fertile neutron capture, the fissile ^{239}Pu. We can then choose the isogenerator mode (i.e., we produce as much plutonium as we consume) or the breeder mode (i.e., we produce more plutonium than we consume). The latter option allows us to increase Pu availability for future new fast reactors. In practice, reactors in isogenerator or breeder mode operate by consuming only ^{238}U, and can adapt to natural or depleted uranium.

If we want eliminate only plutonium stocks, it is also possible to design a concept that allows the burning of plutonium.

If we want to use thorium, we need to begin operation of a reactor where ^{235}U will be the fissile material and thorium the fertile material. This operation will create in the fuel the fissile ^{233}U. After reprocessing and extraction of this ^{233}U, it becomes possible to begin a cycle with a fuel mix of thorium and ^{233}U that consumes only thorium, as well as a cycle with Pu and ^{238}U that consumes only ^{238}U.

Operation of water reactors without spent fuel reprocessing

A number of countries (United States, Sweden) have chosen not to reprocess their fuel. Under these conditions, the only fissile element available is ^{235}U.

Most of the existing reactors are PWRs and boiling water reactors. The use of water as a heat transfer fluid has several consequences at the neutron level.

The hydrogen atoms present in water are excellent for slowing down the neutrons emitted during the fission reaction. These lower-energy neutrons are called “thermal neutrons.” For the same reasons of slowing down and loss of neutron efficiency, these PWR reactors need a minimum of enrichment (3.5%) to be able to operate. Their operation will therefore induce, in the process of enriching natural uranium, an accumulation of depleted uranium, which cannot be used in these reactors. This depleted uranium will be stored as waste, for the moment without any particular use. For example, in France about 300,000 tons of depleted uranium are stored, and every year about 7000 tons are added.

The fuel cycle of this type of reactor is very simple in countries where there is no reprocessing of spent fuel. Uranium is extracted in mines, then enriched, producing the corresponding quantities of depleted uranium. Then, the fuel is fabricated with this enriched uranium, and finally the spent fuel is stored, generally in a water pool to cool it.

The operation of these reactors therefore leads to significant consumption of uranium, accumulation of depleted uranium, and storage of spent fuels, the final fate of which remains under discussion today. In 2020, the Department of Energy (DOE) officially targeted three potential sites to permanently store (in the form of oxide) the 700,000–800,000 tons of depleted uranium now stored (in the form of uranium hexafluoride (UF_6)) in the former enrichment sites of Paducah and of Portsmouth: the Nevada Nuclear Security Site managed by the DOE, the Clive site in Utah belonging to Energy Solutions, and the Andrews site in Texas belonging to Waste Control Specialists.

A PWR producing one GWe for 1 year requires 200 tons of natural uranium for 30 tons of enriched uranium and, therefore, 170 tons of depleted uranium. There are 30 tons of waste on arrival, which contain approximately 1 ton of fission products. These 30 tons, which contain very long-lived waste, are contained in fuel elements, increasing their volume and mass. Safe storage of these elements is not simple because they were not originally designed for this function and because they remain hazardous for long periods of time.

Water reactors with fuel reprocessing

The spent fuel of water reactors still contains around 96% of products usable in a reactor, namely about 95% uranium, (mainly ^{238}U) and 1% plutonium. The rest, less than 4%, corresponds to final waste, with less than 1% made up of other minor actinides, including neptunium, americium, and curium.

In these conditions, certain countries such as France have decided to reprocess their fuel with two objectives. The first is to extract reusable products to reduce uranium consumption. The second is to transform spent fuel into final waste, at a reduced quantity since only 4% of the total mass is concerned. The duration that this waste remains hazardous is also considerably reduced thanks to the extraction of plutonium. This methodology provides a first response to the problem of waste storage. We then arrive at the fuel cycle of the French reactors, in Fig. 1.

On the left line of Fig. 1, we find the timeline presented as mining activities, uranium conversion, uranium enrichment, storage of depleted uranium, manufacture, and use of fuel. However, reprocessing will lead to the creation of 50 tons of final waste per year, consisting of the 4% of fission products and minor actinides. The uranium extracted, which is no longer enriched enough but still more enriched than natural uranium, is returned to the enrichment unit to bring it back to good standards. Plutonium is used to manufacture MOX assemblies containing fissile plutonium in the place of enriched uranium, and these subassemblies are burned in the reactors. All of these operations save around 20% of the annual French uranium consumption, which is around 8000 tons per year.

However, the second passage through the reactor of plutonium and uranium extracted from reprocessing leads to isotopic degradation of uranium and plutonium (with the creation of new isotopes). This degradation makes it very difficult to reuse them in PWR because these new isotopes are neutron poisons, which would disturb the general operation of the reactor.

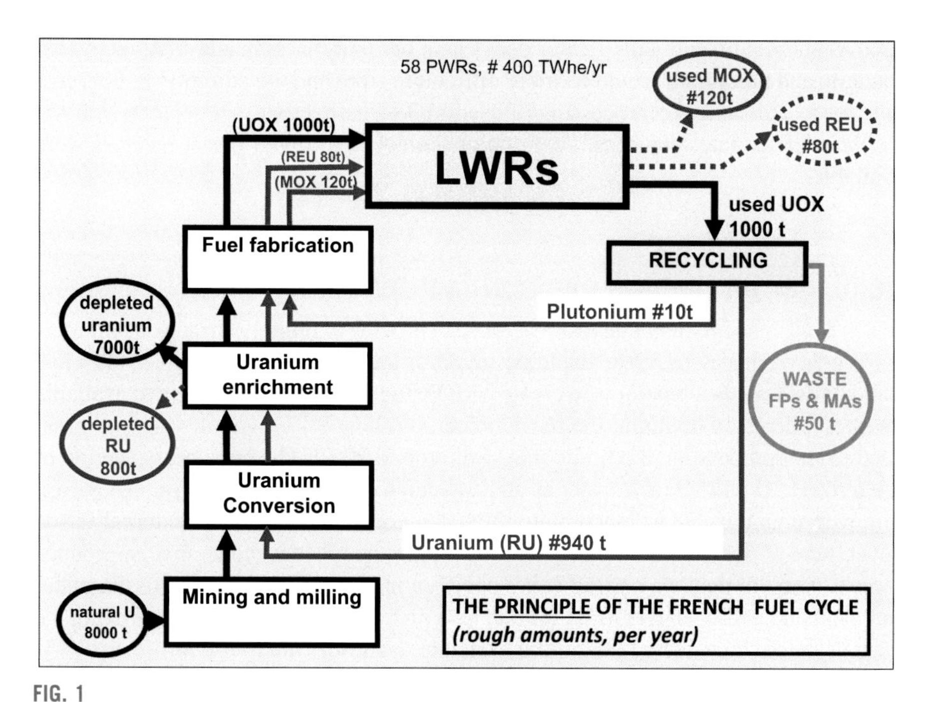

FIG. 1

Fuel cycle of French reactors with reprocessing activities.

We can therefore see that the initial objectives are imperfectly achieved today. We have reused some products with energy value, and we have created waste in a reduced quantity and in a storable and final form. However, certain points remain to be improved upon to reach the initial objectives:

- We have not separated the last minor actinides, which increases the duration these final wastes are hazardous. This point was intensively studied following the French law to organize research on nuclear waste of 1991 (law 91-1381). Studies showed that an effective separation of the last minor actinides (neptunium, americium, and curium) was possible during reprocessing activities. However, this operation was not of great interest because these products are difficult to store and to use in water reactors.
- Plutonium management is not optimized. The use of MOX does not allow the use of all the plutonium produced (approximately 10 tons per year), and the plutonium is re-created in used MOX in a degraded isotopic form (build-up of ^{240}Pu, which is a strong neutron poison). Thus the management of the final plutonium balance is not optimized, and their overall stock continues to increase.

There are still used MOX subassemblies to reprocess. This reprocessing is possible but would lead to the production of final waste, more loaded with actinides such as americium and with greater decay heat per unit mass. In addition, recovered uranium and plutonium would be more difficult to store and are currently in nonreusable form. This makes reprocessing of used MOX unattractive.

We will see later that these three points could be resolved if fast reactors were available.

Reminder on the fuel cycle of fast reactors

It is in cycle management that fast reactors can provide extremely attractive solutions.

When we have stocks of depleted uranium and plutonium available, the most logical fuel for a fast reactor is to create a MOX fuel composed of these two available "wastes": depleted uranium, the fertile material (about 80%), and the plutonium issued from reprocessing, the fissile material (around 20%). The good cross-section of the fertile ^{238}U will allow it to act as an isogenerator reactor, that is, produce by capture with uranium-238 as much plutonium as it consumes. If we add external fertile cover, here of depleted uranium, we can even use the breeder mode, that is, produce more plutonium than we consume. We can then increase the amount of fissile material available, which can be used to load new fast reactors. If, on the other hand, we want to reduce the stock of plutonium available, it suffices not to use fertile type ^{238}U or to have a core design promoting the escape of neutrons.

We then arrive at a simple theoretical fuel cycle shown in Fig. 2 with a fleet of isogenerator fast reactors.

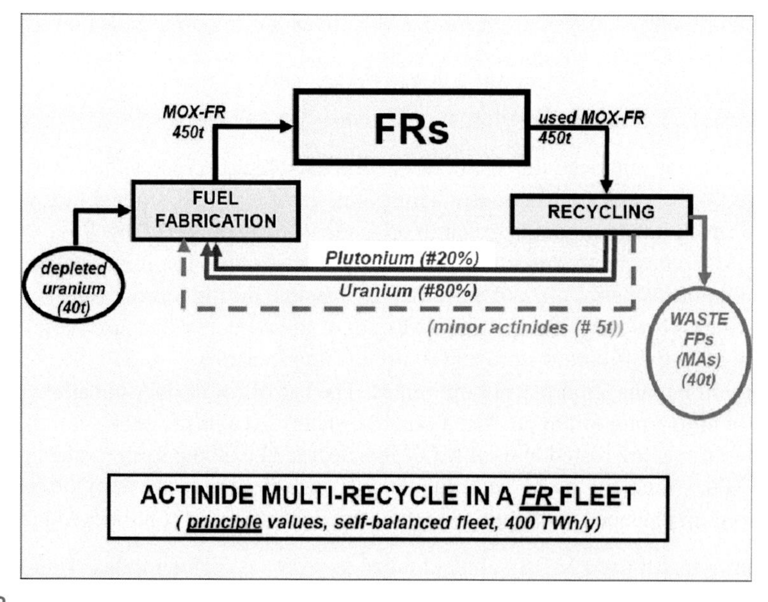

FIG. 2

Theoretical fuel cycle of a fast isogenerator reactor fleet for a production of 400 TWh/year.

We see in this isogenerator cycle that the reactor consumes only depleted uranium and produces a mass of waste almost equivalent to the mass of the initial products.

Note: There is in fact a slight decrease in mass linked to energy production. This decrease can be calculated with Einstein's formula $E = mc^2$, but this value is very small and negligible.

If we choose the breeder option, we will have, after reprocessing, an excess of plutonium available that we can use for a new reactor. For example, on the Phénix fast breeder reactor, which had fertile covers to be breeder, the breeding rate reached measured after fuel reprocessing treatment was 1.15 [1]. If the plutonium burner option is chosen, we need to reintroduce new plutonium during the fuel fabrication.

It is also possible to imagine fast reactors that use different fuels. For example, the Russian sodium fast reactor BN-600 has been in operation for a long time, using enriched uranium fuel. One could also imagine using thorium as a fertile material instead of ^{238}U. India has been exploring this possibility, as it has significant mining possibilities for the production of thorium.

Reactors that no longer need mining and fuel enrichment

Fast reactors consume only depleted uranium for energy production. As the breeder mode makes it possible to gradually obtain the fissile material necessary for loading the next reactor, a fleet of reactors can be formed that consume only depleted uranium.

France, which today has more than 300,000 tons of depleted uranium, and a fleet of reactors of 63 GW, could produce its energy for thousands of years with a fleet of fast reactors using this single stock that is already available.

Uranium mining and enrichment therefore become unnecessary. This point has very positive ecological consequences: a significant decrease in CO_2 emissions due to absence of mining and higher overall electric efficiency due to absence of enrichment. It also removes all geopolitical problems related to the location and availability of uranium mines.

Finally, it presents a source of savings and an important financial advantage.

Almost unlimited energy

The estimates currently made on the conventional energy resources identified in the British Petroleum (BP) statistical review [2] and in the International Atomic Energy Agency (IAEA) (Red Book) report [3] show that the uranium used in its current form in water reactors could produce around 70 gigatons of oil equivalent (Gtoe) under current economic conditions, which makes it a limited contributor to the needs of mankind compared with fossil energy. Fig. 3 shows the potential contribution of uranium compared with that of fossil energy.

On the other hand, if all uranium is burned in fast reactors, this value is multiplied by approximately 100, and the energy obtained from uranium becomes by far the most important energy contributor available, with 7000 Gtoe (all fossil fuels contribute less than 1000 Gtoe) (Fig. 4).

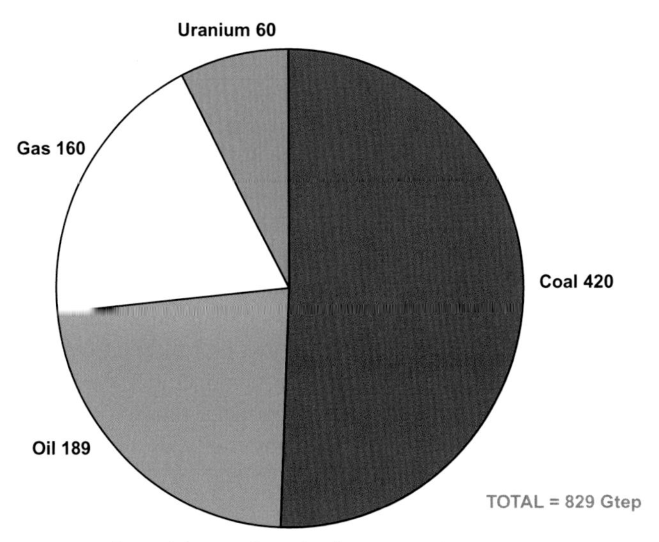

FIG. 3

Contribution of energy sources in different parts of the world. Uranium used in water reactors is indicated in *green*.

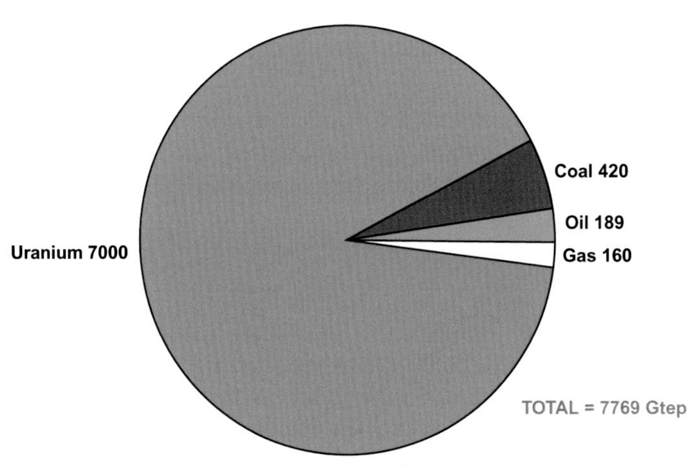

FIG. 4

Contribution of energy sources in different parts of the world. Uranium used in fast reactors is indicated in *green*.

Indeed, so little uranium is required to produce energy that one could easily accept higher mining costs. Given the significant availability of uranium on Earth, we see that this energy joins the peloton of unlimited renewable energies.

Energy that produces neither CO_2 nor dust

Like nuclear power in general, this technology produces no CO_2, neither dust nor air emissions. It is therefore a solution that takes both global warming and air pollution into account.

An energy that has almost zero chemical release in reactor operation

Chemical release related to the operation of water reactors is extremely low. However, certain products relating to water treatment must be rejected as well as gaseous effluents. Comparative analysis with a sodium fast reactor was carried out [4], showing even lower values.

In particular, the necessary containment of liquid sodium means that in operation there is no discharge at this level. Gaseous releases are not measurable because they are too low [4,5].

Operation that minimizes staff dosimetry

Again, the confinement imposed by liquid sodium leads in normal operation to a dosimetry even lower than in water reactors and almost zero for staff and the environment [4,5].

Energy that decreases the global amount of radioactive waste

Depleted uranium and plutonium from PWR subassembly reprocessing are stored with no possible use and become, more or less, waste in the absence of fast breeder reactors.

Fast reactors use this existing and stored waste, which leads to a reduction in their storage quantity.

On the other hand, each reprocessing/remanufacturing cycle will create new metallic waste [6], which must be taken into account in the final assessment of the cycle. However, the activity of this waste is low with short half-life in comparison with that produced by the fuel.

An energy that can drastically reduce the duration final nuclear waste is hazardous

The spent fuels of current reactors contain, before reprocessing, very long-lived waste: uranium, plutonium, minor actinides, etc. The current reprocessing, by recovering plutonium and uranium, already leads to a significant drop in the duration this waste is hazardous, which is reduced by a factor of 10 (Fig. 5).

If we also separate the minor actinides, we arrive at final waste whose half-life is short compared with uranium and minor actinides. In practice, after several hundred years, the residual activity of this final waste is comparable to the activity of the initial mining products (Fig. 5).

Important studies were carried out in France on the separation and the transmutation of these minor actinides following a law of December 30, 1991 (law no 91-1381 known as Bataille law) that organized research on the management of radioactive waste in three axes: geological storage, long-term storage, and separation/transmutation. These studies showed that the separation of minor actinides is feasible with good selectivity.

On the other hand, their transmutation inside a reactor required a fast reactor far more capable of burning this waste than the thermal flux of a water reactor. Many experiments were achieved in the Phénix reactor to demonstrate the potential of fast reactors (see, in [7], Chapter 13 "demonstration of transmutation possibilities," which explains all these experiments).

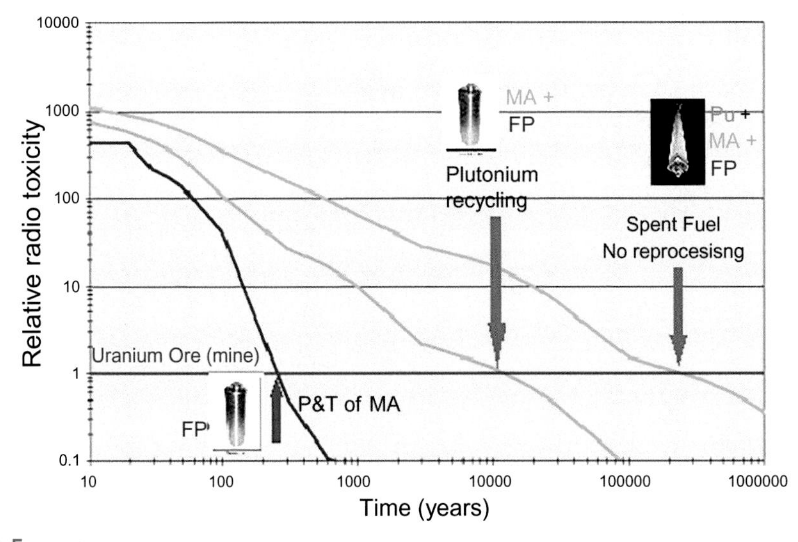

FIG. 5

Decay time of radiotoxicity of waste, without reprocessing, with extraction of plutonium and with extraction of all minor actinides (in the case of partitioning and transmutation of all minor actinides).

The long-term objectives of fast reactors is clearly to burn also the minor actinides issued from reprocessing, that is, americium, curium, and neptunium, to reach this objective of final waste with short duration of radiotoxicity.

A small ecological footprint

Hydro, solar, and wind renewable energies are low-density energies, that is, significant production leads to a strong environmental footprint.

For example, if we compare a fast sodium reactor type ESFR SMART (which will be addressed in further detail in Chapter 5 of this book) with a field farm of wind turbines, we see that, with the latest models, which are more than 200 m high and of 2 GW, it is necessary to cover approximately 2000 km^2 with these wind turbines to obtain a production equivalent to that of a reactor that covers only 1 km^2. The quantities of concrete and steel required for these wind farms are much greater than those required for the reactor, with corresponding consequences in terms of CO_2 production and saving of resources. In addition to clearing and modifying landscapes, operation will cause visual and noise pollution over thousands of square kilometers. Noise keeps fauna away and negatively affects biodiversity, and the turbines lead to a massacre on birds and bats.

In comparison, a reactor, which covers only 1 km^2 and operates without chemical rejection release, without emission of dust or CO_2 and without noise, is clearly the more advantageous technology. When I was director of the Phénix fast reactor, a number of animals, protected because they were endangered, took refuge within the plant, in particular the eyed lizard and the Lulu lark, which made nests in the chimney!

Conclusion

This second chapter outlines the operating principle of fast reactors and the corresponding potential advantages expected. It appears that fast reactors, with reprocessing capabilities, seem to be an interesting solution to the energy problems of humanity.

- Unlimited energy production using only existing and already available fuel such as depleted uranium issued from enrichment activities, and plutonium issued from reprocessing activities.
- No need for uranium mining or enrichment plants.
- Production without CO_2, limiting global warming.
- Production without emission of dust and without chemical discharge during reactor operation.
- Final waste in smaller quantities and with a shorter duration of radiotoxicity, making the waste easier to manage because it becomes harmless in a few hundred years.
- Small ecological footprint compared with the pressure that certain renewable energies exert on the environment.

References

[1] J. Guidez, Chapter 25: Reprocessing and multi recycling/Chapter 13: Demonstration of transmutation possibilities, in: Phenix. The Feedback Experience, EDP Sciences, 2013.
[2] BP, BP Statistical Review 2019, 2019.
[3] IAEA, Red Book 2016, 2016.
[4] J. Guidez, A. Saturnin, Evolution of the collective radiation dose of nuclear reactors from the 2nd through to the 3rd generation and 4th generation sodium-cooled fast reactors, EPJ Nuclear Sci. Technol. 3 (2017) 32, https://doi.org/10.1051/epjn/2017024. published by EDP Sciences, 2017.
[5] J. Guidez, Chapter 23: A positive environmental report, in: Phenix. The Feedback Experience, 2013.
[6] A. Saturnin, J.-F. Milot, C. Chabert, F. Laugier, G. Senentz, B. Carlier, Examining plutonium multi-recycling options for the French reactor fleet in terms of radioactive waste inventory, in: Global 2019, 2019.
[7] J. Guidez, Chapter 13: Demonstration of transmutation possibilities, in: Phenix the Feedback Experience, 2013.

CHAPTER

Fast reactors exist, I've met them

3

Abstract

The Generation IV International Forum (GIF) was created in 2000, to give an international view of all the possible types of Generation IV (Gen-IV) reactors.

This exhaustive review has defined six families of possible reactors.

Two families are dedicated to reactors that are not fast reactors: the very-high-temperature reactor (VHTR) and the supercritical water reactor (SCWR). Therefore, these reactors do not have the advantages of fast reactors and are not analyzed in this book.

The gas-cooled fast reactor (GFR) has been studied very intensively in France after the closure of Superphénix. However, the concept presents major challenges in terms of safety, which is the main reason why a reactor of this type has never been built. In the end, the concept was abandoned.

The lead-cooled fast reactor (LFR) is an interesting case of a fast reactor. Reactors of this type (with eutectic lead/bismuth) have been built and operated by Russia for their submarines. Corrosion effects on reactor material caused this type of reactor to be abandoned. However, Russia continues to study this type of reactor, and the construction of a new reactor BREST has begun. Compared with sodium reactors, the experience feedback remains low, the corrosion problems are not clearly resolved, and other difficulties, including seismic behavior, exist. Therefore, this family of reactor has not been studied further in this book.

There remain two families of fast reactor:

- The sodium-cooled fast reactor (SFR), which has considerable experience feedback, will be the object of the three following chapters.
- The molten salt reactor (MSR), which has little experience feedback but seems to present very interesting possibilities for the future, will be the object of the Chapter 7.

Keywords

Generation four, GIF, SFR, MSR, VHTR, SCWR, LFR, GFR

Fast Reactors. https://doi.org/10.1016/B978-0-12-821946-1.00002-4

Introduction

In January 2000, the US Department of Energy's Office of Nuclear Energy, Science and Technology convened a group of senior governmental representatives from nine member countries to begin discussions on international collaboration in the development of nuclear energy systems for the future, called Gen-IV.

This group, subsequently named the GIF Policy Group, also decided to form a group of senior technical experts to explore areas of mutual interest and make recommendations regarding both research and development areas, as well as processes by which collaboration could be conducted and assessed. This senior Technical Experts Group first met in April 2000.

The founding document of the GIF is set out in the GIF Charter, first signed in July 2001 by Argentina, Brazil, Canada, France, Japan, the Republic of Korea, South Africa, the United Kingdom, and the United States. The Charter has since been signed by Switzerland (2002), Euratom (2003), and most recently the People's Republic of China and the Russian Federation in November 2006. Australia joined GIF in December 2017, and the United Kingdom become an "active member" by ratifying the Framework Agreement in October 2019.

For more than two decades, GIF has led international collaborative efforts to develop next-generation nuclear energy systems that can help meet the world's future energy needs. Gen-IV goals aim to use fuel more efficiently, reduce waste production, be economically competitive, and meet stringent standards of safety and proliferation resistance.

With these goals in mind, some 100 experts evaluated 130 reactor concepts before GIF selected 6 reactor technologies for further research and development. These concepts include: gas-cooled fast reactor (GFR), lead-cooled fast reactor (LFR), molten salt reactor (MSR), supercritical water-cooled reactor (SCWR), sodium-cooled fast reactor (SFR), and very-high-temperature reactor (VHTR).

Information on the status in 2021 of GIF system arrangements and memoranda of understanding is shown in Fig. 1 below.

SFR			•	•	•	•	•			•	•	•
VHTR	•	•	•	•	•	•			•	•	•	•
LFR			•		•	•	•			•		•
SCWR		•	•		•		•					•
GFR				•	•							•
MSR	•	•		•			•		•	•		•

• : signatory of System Arrangement
• : signatory of Project Arrangement • : signatory of MoU

FIG. 1

Status of the six-reactor technology and of the countries involved in these technologies.

Two of these candidates are not fast reactors. They do not make it possible to achieve the objectives of renewable energy in a circular fashion with a closed cycle. They will therefore be eliminated from our analysis.

The very-high-temperature reactor

High-temperature gas-cooled reactors (HTGRs or simply HTRs) are helium-cooled graphite-moderated nuclear fission reactors utilizing fully ceramic fuel (Fig. 2). They are characterized by inherent safety features, excellent fission-product retention in the fuel, and high-temperature operation suitable for the delivery of industrial process heat. Typical reactor-coolant outlet temperatures range between 750°C and 850°C, thus enabling power-conversion efficiencies up to 48%. The VHTR is understood to be a longer-term evolution of the HTR, targeting even higher efficiency and more versatile use by further increasing the helium outlet temperature to 950°C or even higher, up to 1000°C. However, such high temperatures would require the use of new structural materials not available today.

These reactors have a very specific fuel. They can be in the form of balls or prisms, but in all cases, they consist of highly enriched uranium (less than 20%) within a carbon matrix, which plays the role of moderator. They are therefore thermal reactors. The quantity of uranium-235 needed to produce a kilowatt is greater than that of a pressurized water reactor (PWR), despite very high burnup rates, because of the initial enrichment.

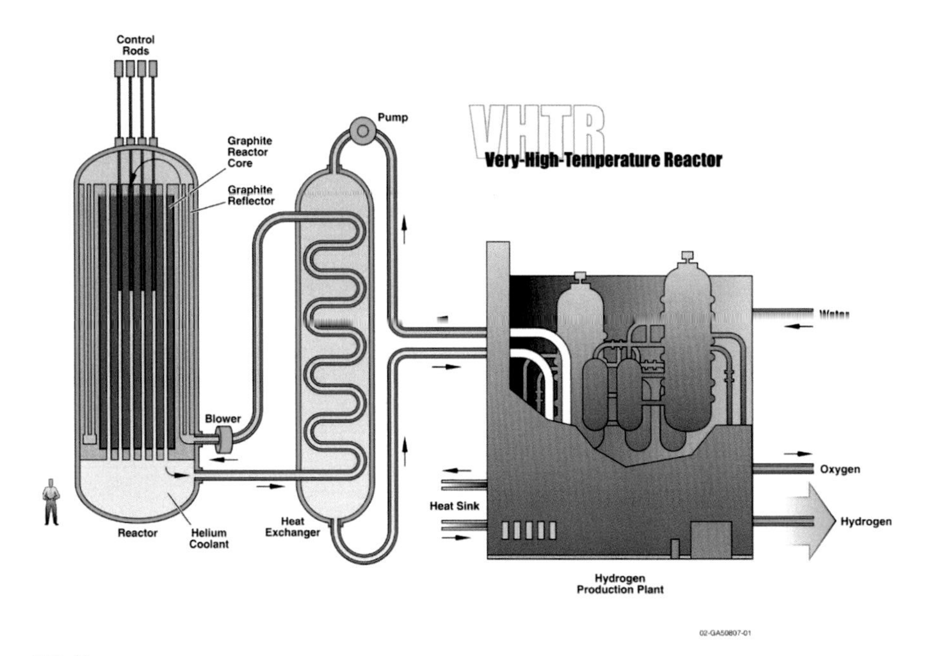

FIG. 2

View of VHTR principle.

This fuel is not initially designed for reprocessing, but as ultimate waste after combustion. Reprocessing tests revealed that it is necessary to grind the carbon for elimination before attempting to reprocess. These tests show that this elimination is complex or even impossible, in particular because of the carbon-14 elimination problem. We are therefore in a cycle that is not closed, with very long-lived waste and without the aspect of circular economy of the fast reactors.

Finally, these reactors, several prototypes of which were built in the 1980s, were abandoned for economic reasons, which still apply today. Their very low power density leads them to have much larger dimensions than other reactors. The 100 MWe prototype being started up in 2022 in China has a tank of much larger dimensions than a 1600 MWe PWR.

Finally, in terms of safety, these reactors are presented as very safe in a passive way, because if we lose the coolant and if we do nothing, the fuel slowly rises in temperature to peak at around 1600°C before stabilizing and then slowly descending. This is only valid for low-power reactors, around 100 MWe. In addition, restarting a reactor after reaching such temperatures is certainly problematic for its structures. However, the safety guidelines have changed, especially for plane crashes and related fires. Carbon that rises in temperature as a result of the loss of the cooling gas should not ignite. Protection against aircraft crashes would lead, given the size of these reactors, to prohibitively expensive containment. More generally, the basic safety guideline to be observed for the reactors of the future is to have no incident leading to radioactive releases outside the site. This brings us back to the problem of the quality of the containment of these large reactors.

To conclude, without presenting here an exhaustive review of the advantages/disadvantages of this type of reactor, it is clear that it does not present any of the advantages promised by fast reactors. It consumes more ^{235}U than PWRs and produces long-lived waste in an open cycle.

This type of reactor is therefore excluded from our analysis.

The supercritical water reactor

SCWRs are a class of high-temperature, high-pressure water-cooled reactor operating with a direct energy-conversion cycle and above the thermodynamic critical point of water (374°C and 22.1 MPa) (Fig. 3).

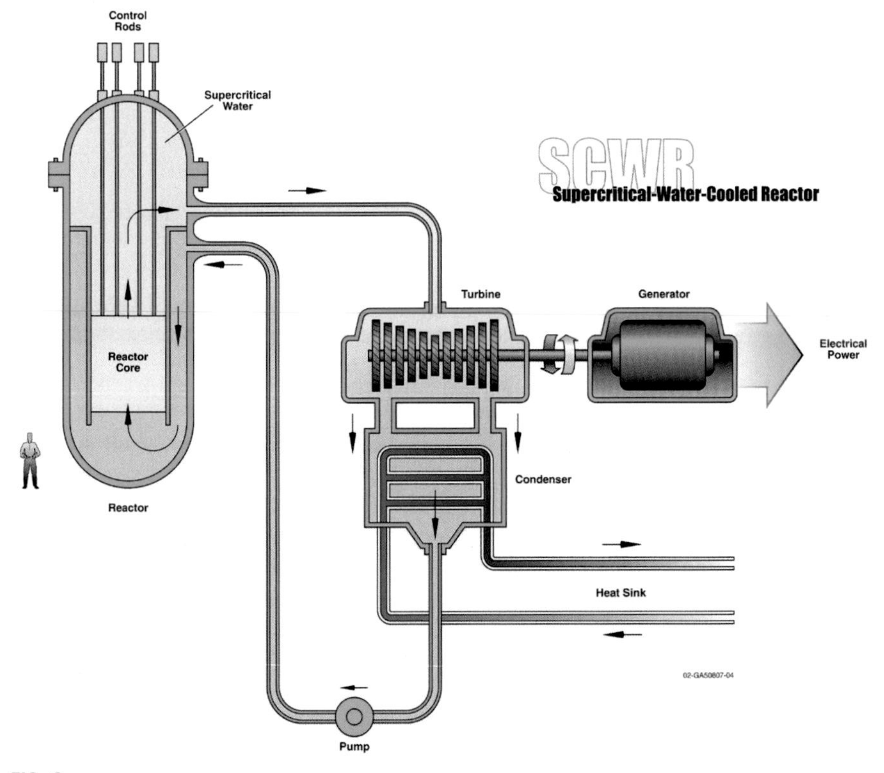

FIG. 3

View of SCWR principle.

When water is in the supercritical state, its physical properties change enormously, which allows significant gains in the compactness of the core and the reactor. For example, its heat capacity is increased by a factor of six. The neutron flux is then slightly faster than on a PWR, but remains a thermal flux since the cooling fluid remains water.

The operating water pressure is around 240 bars, which of course does not go in the direction of improving safety and would not be possible for the sizing of a vessel. In practice, only Canadian CANDU-type reactors could possibly withstand this pressure increase. In a CANDU reactor, each fuel is inside a pressure tube of a relatively small diameter and whose increase in thickness, to support the pressure increase from 140 to 240 bars, would still be sizable.

Note that the principle of these reactors poses additional safety problems that do not go in the direction of improving the safety of the concepts. For example, in the event of depressurization accidents, water suddenly regains its usual properties, which leads to complex transient situations. Likewise, note that the water entering the fuel element is under pressure but not in its supercritical form. It only becomes supercritical when the temperature rises in the assembly. This leads to quite complex neutron stability problems inside the core.

In conclusion, this type of reactor does not present any of the advantages promised with fast reactors and aims only for economic gains in the water sector. It consumes as much uranium as current water reactors and does not present any innovation to reduce waste or for a closed cycle.

This type of reactor is therefore excluded from our subsequent analysis.

At this stage, therefore, only four families remain, all of which are fast reactors. However, the author will rule out two of them, for reasons that perhaps seem clear to him alone, but that are nevertheless shared by a large number of specialists.

The gas-cooled fast reactor

The GFR is a high-temperature helium-cooled fast-spectrum reactor with a closed fuel cycle (Fig. 4). The core outlet temperature is in the order of 850°C. The GFR combines potentially the advantages of fast-spectrum systems for the long-term sustainability of uranium resources and waste minimization (through multiple fuel reprocessing and the fission of long-lived actinides) with those of high-temperature systems (e.g., high thermal-cycle efficiency and industrial use of generated heat).

This concept was particularly studied in France, at the French Atomic Energy Commission (CEA), after the closure of Superphénix. The general idea was to keep the advantages of the fast process but in a concept where water was removed and where certain potential advantages could arise thanks to the high temperatures.

It appeared from the beginning of the studies that the oxide fuel was not sufficiently conductive to ensure, with a gas, the power generation. Carbide fuel was needed.

Many other difficulties arose (materials, radiation, etc.), but above all the design of the reactor suffered from there having never been any reactors of this type built,

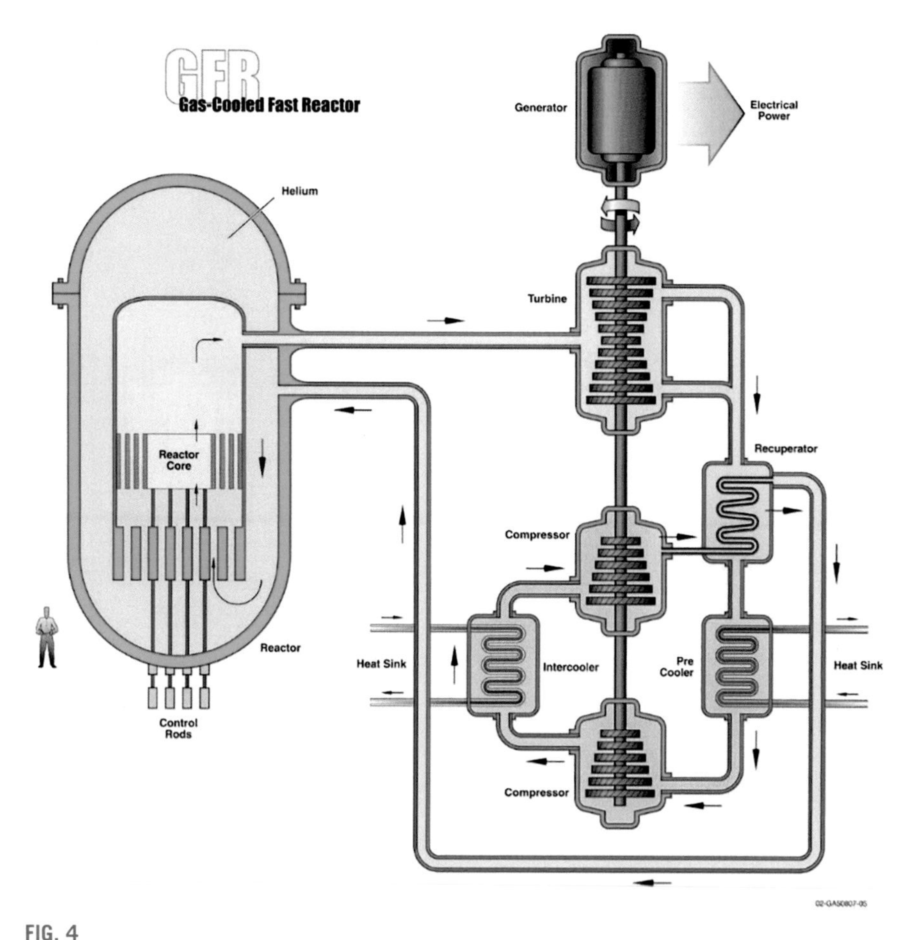

FIG. 4

View of GFR principle.

and because of this there probably never will be. If the cooling flow is stopped, the absence of a moderator and the high power density lead to a considerable rise in temperature. However, the probability of a cooling shutdown occurring is not negligible, especially with a helium-type gas. Workarounds have been sought, but nothing very convincing has emerged on which to build a convincing safety analysis.

The studies were abandoned after a few years, and currently there are no more "serious" teams working on the subject in the world.

We will therefore not return to this concept in our subsequent analysis.

The lead-cooled fast reactor

The LFR (Fig. 5) is characterized by a fast-neutron spectrum, a closed fuel cycle with full actinide recycling, and high-temperature operation at low pressure. The coolant

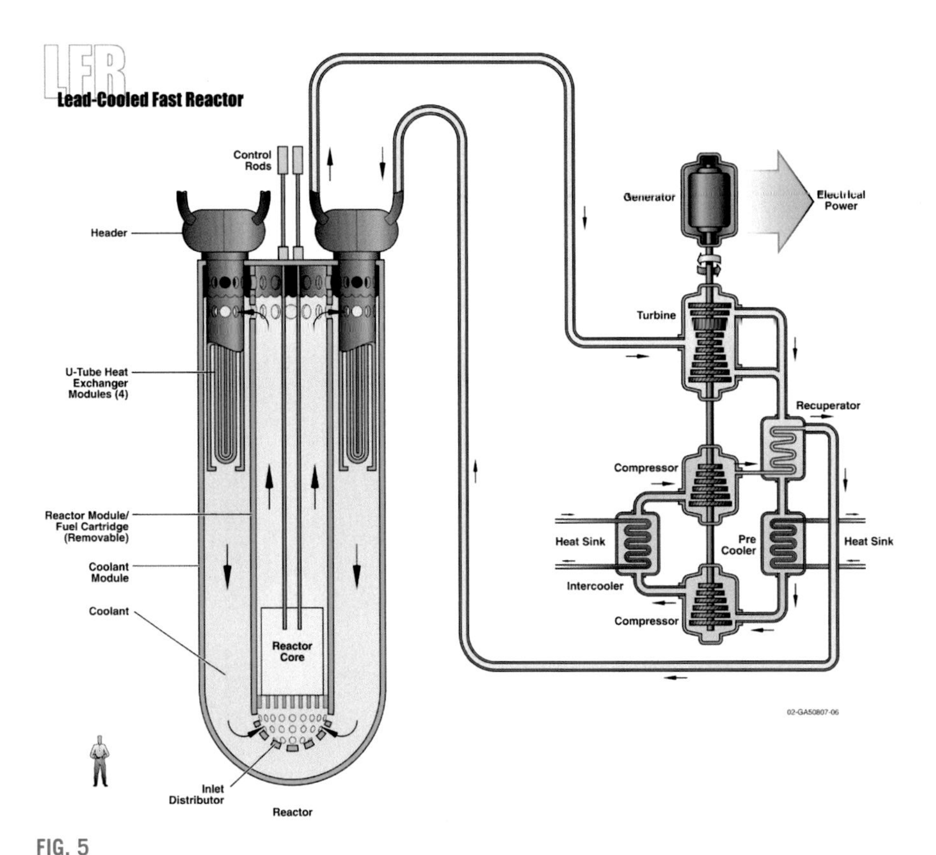

FIG. 5

View of LFR principle.

may be either lead (preferred option) or lead-bismuth eutectic (LBE). The LFR may be operated as a breeder or as a burner of actinides from spent fuel, and it may use inert matrix fuel or act as a burner/breeder using thorium matrices.

Lead and LBE are chemically inert liquids with very good thermodynamic properties. The LFR could have multiple applications, including the production of process heat.

An important feature of the LFR is the enhanced safety that results from the choice of lead as a chemically inert and low-pressure coolant. In terms of sustainability, lead is abundant and hence available, even in the case of the deployment of a large number of reactors. More importantly, as with other fast systems, fuel sustainability is greatly enhanced by the conversion capabilities of the LFR fuel cycle. Because they incorporate a liquid coolant with a very high margin to boiling and limited interaction with air or water, LFR concepts offer substantial potential in terms of safety.

Lead reactors have a history. After the Second World War, during the Cold War, the United States and Russia wanted to develop nuclear submarines. After testing different types of reactors, the United States finally chose to develop a PWR, the ancestor of all future PWRs. The Russians chose a fast lead reactor, which has the particular advantage of good neutron protection in a confined space. This reactor, in fact, used LBE, which lowers the melting point from 327°C to 125°C, making the operation of the reactor easier.

Unfortunately, lead is an extremely corrosive material, particularly toward nickel and iron. To protect the reactor materials, the operator injects oxygen to create a protective oxide layer. This is a delicate operation because, if we do not inject enough, corrosion occurs, and if we inject too much, we create a powder of insoluble lead oxides capable of plugging the channels of the core. In addition, this protection is effective only if it remains below 550°C, if there is no local cavitation likely to remove the oxide layer, and if the circulation speed remains low, to avoid removing this layer of oxides. For example, this was impossible for mechanical pumps where the impellers had to be ceramic. This also had consequences for the reactor core where the low speeds required greater needle spacing, which required a nitride fuel. Despite all these precautions, problems have arisen. At least one submarine has gone missing, and another has returned to harbor with the crew above the ship. Ultimately, this type of reactor was abandoned for Russian nuclear submarines.

However, Russia continues to work on this concept. At the material level, research continues to seek a material that does not depend on an oxide layer to resist corrosion by lead. Materials loaded with silicon or aluminum have been tested. Coating experiments with aluminum have yielded good results.

In this context, Russia announced the construction of a 300 MWe BREST-300 lead reactor, for which the first concrete was poured in 2021. Nitride fuels were manufactured and are being tested in the BN-600 and BN-800 reactors, for application on BREST. At the global level, among other endeavors, China has created a research center on lead reactors, research is being carried out in Europe (Alfred project, materials studies in Karlsruhe, etc.), and Westinghouse has announced a lead small modular reactor (SMR) project.

Unlike GFRs or SCWRs, we cannot say that lead reactors have no future.

However, even if the material problems find a solution, a number of problems remain. Some are “minor” because they should be possible to resolve, including the production of highly toxic polonium, the impossibility of washing the components, the design of more complex control rods (they float in the lead), and the very high pumping power necessary because of high density.

Others are more fundamental:

- Resistance to earthquakes is very difficult for this type of reactor because of the density of lead, in particular on the horizontal plane of the vessel. This seems to condemn this type of reactors to small powers.
- Limiting the temperature to 550°C (for corrosion reasons) prevents full benefit of the advantages of the high vaporization point of lead. On the other hand, the melting point at 327°C results in a very low temperature range. This precludes the secondary circuit option because of the risk of lead freezing in some situations. This is an economic advantage because these circuits are expensive. However, this is a challenge in terms of safety because the steam generators

are then located directly in the primary tank with the risk of water/liquid-metal interaction inside this primary vessel during operation of the plant.

We then arrive at the somewhat complex drawing of a small reactor with a complex vessel in a concrete well (Fig. 6).

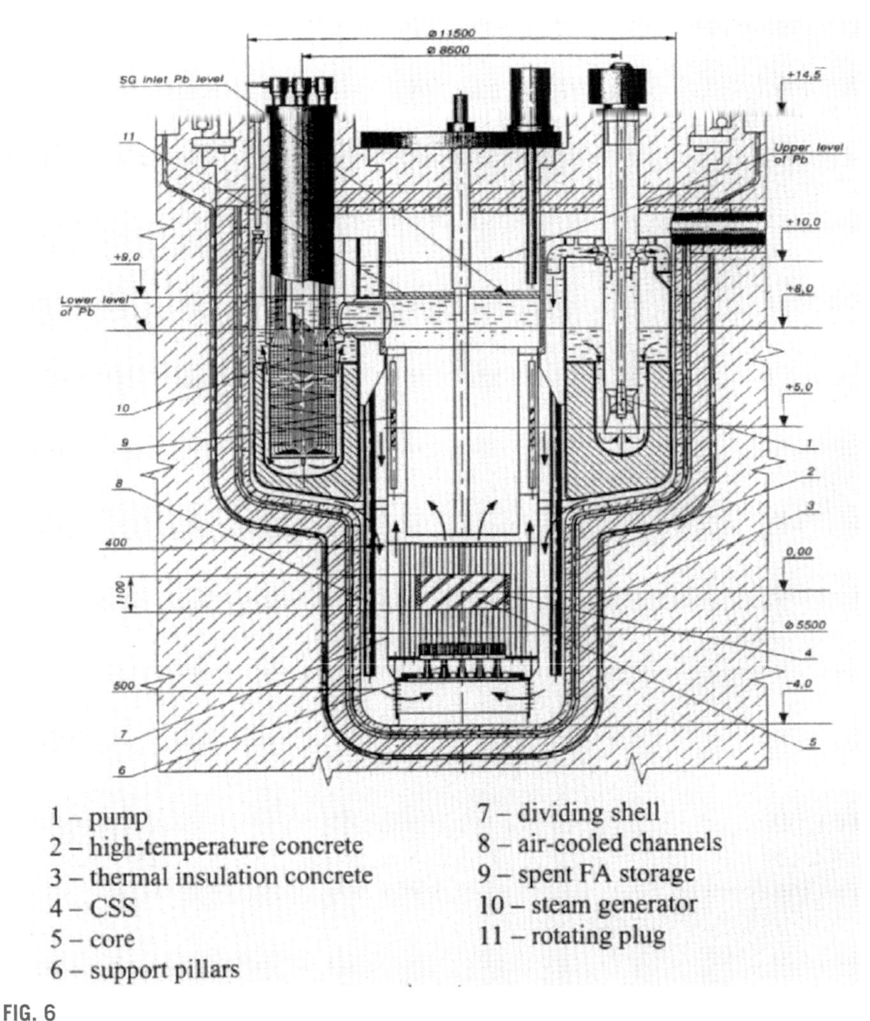

FIG. 6

Drawing of the BREST reactor.

If the problem of material resistance is resolved and if BREST works well, it is possible that these reactors have a future, especially in the context of small SMRs where they benefit from their intrinsic neutron protection.

This being the case, we will not take the analysis further in this sector.

There remain the two most promising sectors to analyze: SFRs and MSRs.

The sodium-cooled fast reactor

The SFR (Fig. 7) uses liquid sodium as the reactor coolant. It features a closed fuel cycle for fuel breeding and/or actinide management. A variety of fuel options are being considered for the SFR, with mixed oxide preferred for advanced aqueous recycle and mixed metal alloy preferred for pyrometallurgical processing. Owing to the significant amount of experience accumulated with sodium-cooled reactors in several countries, the deployment of SFR systems is already feasible.

Using liquid sodium as the reactor coolant allows high power density with low coolant volume fraction and operation at low pressure. While the oxygen-free environment prevents corrosion, sodium reacts chemically with air and water and requires a sealed coolant system.

Plant size options under consideration range from small (50–300 MW_{el}) modular reactors to larger plants of up to 1500 MW_{el}. The outlet temperature is 500–550°C for these options, which allows for the use of materials developed and proven in prior fast-reactor programs.

The SFR closed fuel cycle enables regeneration of fissile fuel and facilitates management of minor actinides. However, this requires that recycle fuels be developed and qualified for use. Important safety features of this Gen-IV system include a long thermal response time, a reasonable margin to coolant boiling,

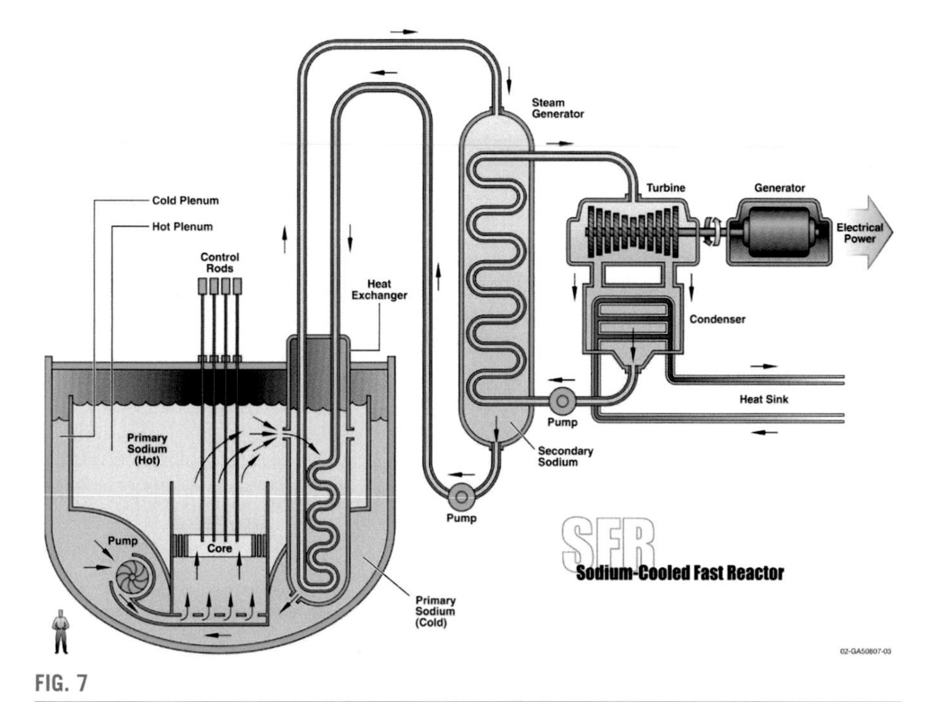

FIG. 7

View of SFR principle.

a primary system that operates near atmospheric pressure, and an intermediate sodium system between the radioactive sodium in the primary system and the power-conversion system. With innovations to reduce capital cost, the SFR is aiming to be economically competitive in future electricity markets. In addition, the fast-neutron spectrum greatly extends the uranium resources compared with thermal reactors. The SFR is considered to be the nearest-term deployable system for actinide management.

Much of the basic technology of the SFR has been established in former fast-reactor programs and is today confirmed by the BN-600 and the BN-800 operation in Russia. New programs involving the SFR technology include the Chinese experimental fast reactor, which was connected to the grid in July 2011, two Chinese reactors of 600 MWe under construction, and India's prototype fast breeder reactor.

The SFR is an attractive energy source for nations that would like to make the best use of limited nuclear fuel resources and manage nuclear waste by closing the fuel cycle.

Chapter 5 is dedicated to the most advanced SFR concept developed in Europe: the ESFR SMART, based on various European SFR experiences, including PFR, Rapsodie, Phénix, Superphénix, EFR, and more recently the ASTRID project.

The molten salt reactor

MSR concepts have been studied since the early 1950s, but with only one demonstration reactor (Aircraft Reactor Experiment, ARE) and then one prototype reactor (Molten-Salt Reactor Experiment, MSRE) operated at the Oak Ridge National Laboratory (ORNL, United States) in the 1960s.

MSRs use molten salts as coolant with fuel dissolved in this coolant. In the MSR, the fission occurs directly in the carrier salt, which transfers its heat to another coolant salt in the heat exchangers (Fig. 8). MSRs are operated at low pressure, slightly above atmospheric pressure. The MSR is a concept and not yet a technology. Indeed, the MSR generic name covers thermal or fast reactors, operated with a U/Pu or a Th/^{233}U fuel cycle, or as transuranium burners, with a fluoride or a chloride salt. Depending on the fuel cycle, MSRs can reuse fissile and fertile materials from light-water reactors. As fissile material, they can also use uranium or plutonium, or minor actinides.

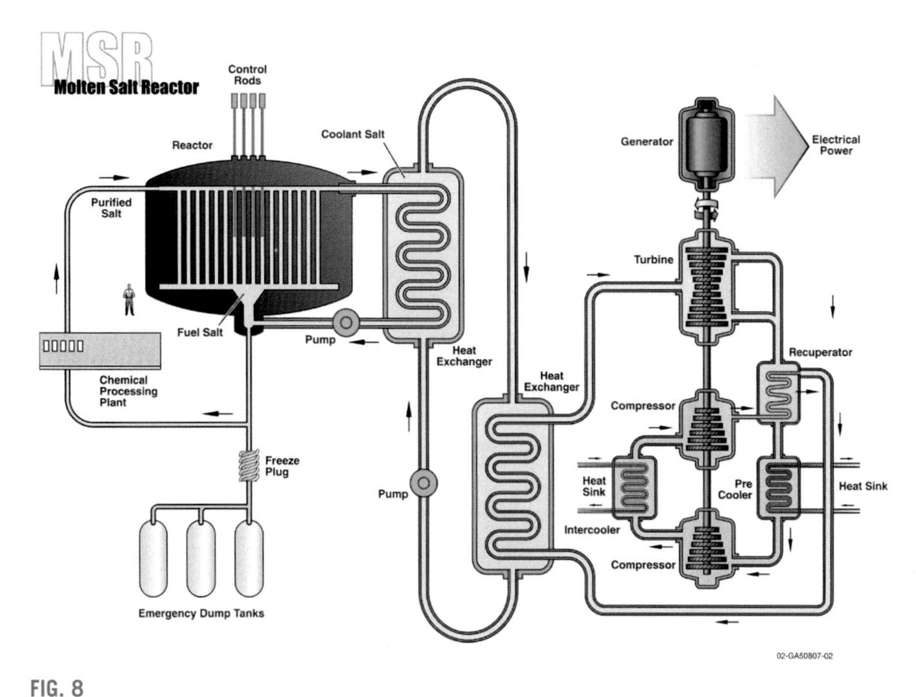

FIG. 8

View of MSR principle.

In the past years, there has been a renewal of interest in this reactor technology, in particular because of its acknowledged inherent reactor safety and its flexibility. The most studied concept is depicted in Fig. 9.

MSRs can be deployed as large power reactors or as SMRs. Today, their deployment is limited by technological challenges, including high temperatures, structural materials, and corrosion. These technical challenges will lead in the next years to the construction of demonstration reactors, some of which is already underway.

In Chapter 6, we will return to the analysis of this concept, which has significant potential advantages, in particular in its fast-reactor version.

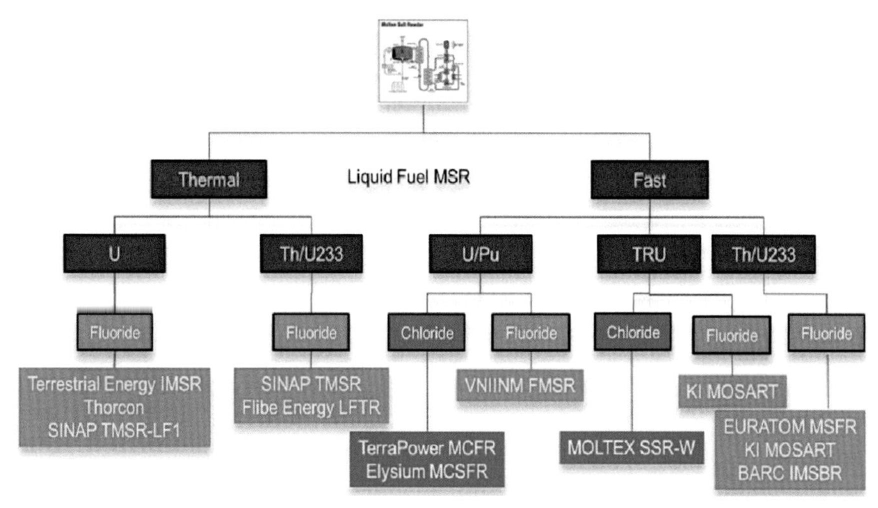

FIG. 9

The most studied MSR concepts, with key players.

Conclusions

The Generation IV International Forum made it possible to analyze all (more than 130) imaginable concepts for the reactors of the future. This forum deduced six possible reactor concepts.

Two concepts are not fast reactors and therefore do not have their advantages (VHTR and SCWR).

Two fast reactors, GFR and LFR, seem a priori without a significant future. The GFR has safety concerns, which caused its study to be abandoned. The LFR is limited by problems regarding corrosion of the materials and earthquake resistance standards. However, Russia continues to show interest in SMR with the construction of BREST.

There remain, in all these concepts of reactors of the future, two concepts that seem promising in that they allow us to benefit from the advantages of fast reactors, set out in Chapter 2.

- The sodium fast reactor, which has by far the most significant experience feedback with more than 20 reactors built in the world, with reactors under construction, and with two power reactors in operation in 2022: BN-600 and BN-800 in Russia. In Chapters 4–6, we will analyze the future of these reactors.
- The fast version of MSR, which has significant theoretical advantages in terms of safety, fuel cycle, and acceptability. However, the experience feedback from this sector is limited, and the feasibility remains to be determined by the construction and operation of demonstration reactors. In Chapter 7, we present an analysis of this type of reactor.

CHAPTER

4 Analysis of the reasons for the failure to deploy sodium-cooled fast reactors globally in 2021

Abstract

The potential advantages of fast reactors were quickly identified, and most nuclear countries started building sodium-cooled fast reactors in the 1950s. However, in 2021, water reactors constitute the bulk of reactors in the world.

We review the 20 or so sodium reactors that have existed or are in operation. Russia, with two power reactors in operation, China, with two reactors under construction, and India, starting up a new sodium reactor, are currently the main countries remaining actively involved in this sector.

This deployment failure is due to a number of reasons discussed in this chapter. These are mainly the availability of low-cost uranium, the technical difficulties inherent in prototype reactors, the higher cost of the fast reactor with its fuel fabrication factory, and the need for operational reprocessing plants to close the fuel cycle.

The proliferation problems attributed to plutonium led the United States not to reprocess its fuel, which led it to lose interest in and abandon its fast reactor program. Few countries now have the necessary reprocessing plants to take advantage of the possibilities of the fast reactor process.

In conclusion, establishing fast reactors, together with the required reprocessing plants and fuel fabrication factories, requires a continuous political and financial effort over a long period, with benefits that will become apparent only in the long term.

Keywords

Uranium availability, Nonproliferation, Waste, Reprocessing, Fuel fabrication

Fast Reactors. https://doi.org/10.1016/B978-0-12-821946-1.00006-1

Assessment of sodium-cooled fast reactors in the world in 2021

After the Second World War, various reactor types were studied and built to guide future strategies for the civilian exploitation of nuclear fission energy. In this context, the potential advantages of fast reactors were quickly highlighted. In addition, the use of liquid metals made it possible to manufacture a reactor without pressure and with good efficiency.

The first reactor in the world to produce current was a fast reactor using a sodium/potassium eutectic as coolant, in 1951. The nuclear countries of the time very quickly built prototypes of fast liquid metal reactors and converged on the use of sodium. England, Germany, France, the United States, and Russia all embarked on this adventure. Later, Asian countries (India, China, Japan, South Korea) took over projects and constructions.

Table 1 [1] summarizes all sodium-cooled fast reactors built (or under construction) around the world, providing very significant operating feedback, of several hundred years.

The United States, Germany, and England shut down their reactors and projects in this area from 1991 to 1994.

In Europe, only France had continued with the Superphénix reactor, after the Rapsodie and Phénix reactors. A new project called ASTRID was initiated in 2010 and then abandoned. The shutdown of Superphénix in 1997, then the shutdown of the ASTRID project in 2020, left France and Europe in 2021 without a constructive project in this area.

Japan, which had a JOYO prototype reactor and the first MONJU power reactor, also started construction of a reprocessing plant in Rokkasho. In 2021, following the 2011 tsunami that caused the Fukushima-Daiichi accident, MONJU was permanently closed, JOYO was shut down, awaiting authorization to restart, and the Rokkasho plant has not yet started up.

In India, filling the Prototype Fast Breeder Reactor (PFBR) reactor with sodium has posed technical problems since several years.

In Russia, after the shutdown of reactors cooled by a lead/bismuth eutectic for nuclear submarines, many prototype sodium reactors were built until the success of the BN-600 reactor. This success allowed the construction and recent startup of the BN-800 reactor, but the following project BN-1200 is currently frozen (Fig. 1). Meanwhile, the fast lead reactor project BREST (300 MWe) will enter the construction phase.

Finally, China, with the help of Russia, has recently started, after the construction of a prototype, the construction of two 600 MWe reactors.

Table 2 presents all the fast reactors in the world either in operation or awaiting operation at the end of 2021. Note that BOR 60 is at the end of its life (it will be replaced by the MBIR under construction) and that the JOYO experimental reactor is awaiting authorization for restart. For the Indian PBFR, the start of the reactor has been causing problems since the beginning of filling operation and the date is uncertain.

Table 1 Fast neutron reactors—Historical and current status.

Electrical power of plant (MWe)	Thermal power (MW)		Dates of plant operation
United States			
EBR 1	0.2	1.4	1951–63
EBR II (E)	20	62.5	1963–94
Fermi 1 (E)	61	200	1963–75
SEFOR		20	1969–72
Fast Flux Test Facility (E)		400	1980–93
United Kingdom			
Dounreay FR (E)	15	65	1959–77
Prototype FR (D)	250	650	1974–94
France			
Rapsodie (E)		40	1967–83
Phénix (D)	250	563	1973–2009
Superphénix (C)	1240	3000	1985–98
Germany			
KNK 2 (E)	20	58	1972–91
India			
FBTR (E)	13	40	1985–2030
PFBR (D)	500	1250	Under filling with sodium
Japan			
JOYO (E)		50, 75, 140	1978–2007
2022 restart?			
Monju (D)	280	714	1994–96, 2010
Kazakhstan			
BN-350 (D)	135	750	1972–99
Russia			
BR 1/2 (E)		1/0.1	1955, 1956
BR 5 Obninsk (R)		5	1958–71
BOR 60 Dimitrovgrad (R)	12	60	1969–?
BR 10 Obninsk (R)		8	1973–2002
BN-600 Beloyarsk 3 (D)	600	1470	1980–?
BN-800 Beloyarsk 4 (E)	864	2100	2017–?
MBIR (E)	40	150	Under construction
China			
CEFR (E)	20	65	2010–?
CFR600 (D)	600	1500	Under construction

C, commercial; D, demonstration or prototype; E, experimental; R, research.

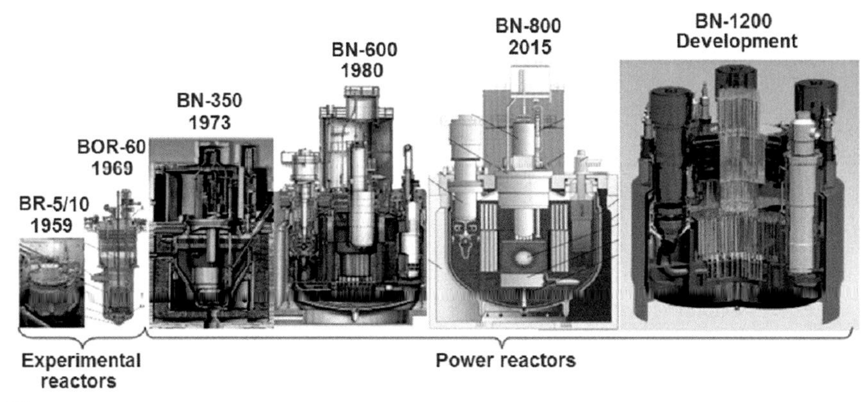

FIG. 1

Development of the rapid sodium sector in Russia.

Table 2 Overview of sodium-cooled fast reactors in operation around the world at the end of 2021.

Reactor	Type, coolant	Power, MW thermal/ electric	Fuel (future)	Country	Notes
BOR-60	Experimental, loop, sodium	55/10	Oxide	Russia	1969–2020s
BN-600	Demonstration, pool, sodium	1470/600	Oxide	Russia	1980–?
BN-800	Demonstration, pool, sodium	2100/864	Oxide	Russia	2017–?
FBTR	Experimental, pool, sodium	40/13	Carbide (metal)	India	1985–2030
PFBR	Demonstration, pool, sodium	1250/500	Oxide (metal)	India	(2022?)
CEFR	Experimental, pool, sodium	65/20	Oxide	China	2010–?
JOYO	Experimental, loop, sodium	140/–	Oxide	Japan	1978–2007, maybe restart 2022

In conclusion, the deployment of sodium-cooled fast reactors, which was very promising in the 1950s, proved to be a failure in 2021, compared with the deployment of water reactors, which are today the dominant reactors on the planet. Only Russia currently keeps two power reactors in operation, and only China (with Russian help) and India currently have reactors under construction.

What are the reasons for this failure, despite the initially recognized potential of this sector?

Technical difficulties specific to the sodium sector?

Some prototypes have experienced considerable technical difficulties, including the English PFR (Prototype Fast Reactor) reactor, and the first Russian reactors, such as the BN-350 reactor, where very significant sodium-water reactions occurred in the steam generators. The operation of Phénix and Superphénix will also be punctuated by various incidents, which will require corrective interventions (see Refs. [2, 3]).

As all incidents that have occurred have been reported [4,5], it is now possible to manufacture and operate these reactors on the basis of this experience feedback. Many incidents that occurred at Phénix (clad failures, sodium leaks due to thermal striping, exchanger ruptures due to a design defect, etc.) were avoided on Superphénix, thanks to this feedback (see Chapter 27 of Ref. [3]). After a difficult start (leak from the used fuel sodium storage tank, sodium pollution by an air inlet) and improvements due to advancements in knowledge (taking into account sodium spray fires), the Superphénix reactor worked very well before a new government allied with the antinuclear parties ordered it to be shut down [3].

The case of the Russian BN-600 reactor (600 MWe) is exemplary in this regard. After some youthful failures at startup, this reactor went on for years of operation with availability rates of around 80%, comparable to those of pressurized water reactors (PWRs) (Fig. 2).

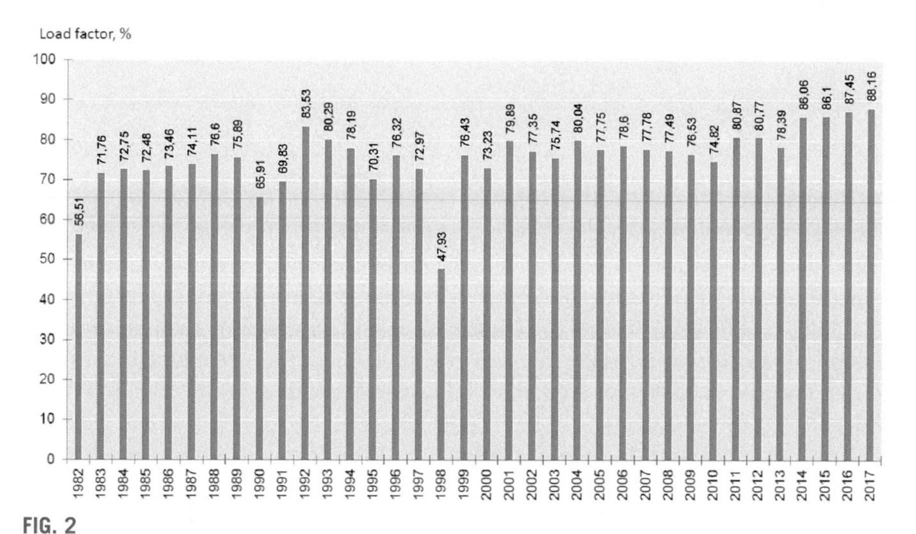

FIG. 2

BN-600 reactor load factor.

It is this success that enabled the construction of BN-800 (800 MWe), which has been operating successfully since its start in 2017 (Fig. 3).

With the experience feedback (REX) available, we can now reasonably say that the construction and operation of these reactors is possible, with a correct load factor, as long as this REX is taken into account (see Appendix 1).

FIG. 3

View of the reactor hall of the BN-800 reactor.

In addition, it should be remembered that water reactors also had many known technical difficulties that they have gradually overcome. In fact, any new industry must resolve a number of problems before it can achieve optimal operating conditions.

The additional cost compared with PWRs?

The design of sodium-cooled fast reactors leads to additional costs compared with PWRs during construction. In particular, the existence of the secondary circuit, necessary to avoid risk of sodium-water reaction in the primary circuit, induces significant additional costs. Attempts to estimate these surcharges have been made in the context of studies on the European fast reactor (EFR). They concluded that there was an intrinsic extra cost of around 30% for reactors built in series. This is a fairly widely accepted value today.

However, we must add to this additional cost the prototype effect, which automatically leads to other additional costs compared with reactors built in series. It is indeed necessary to develop specific industrial production lines, which is very expensive and pays for itself only after successive orders.

For example, in the case of Superphénix, these two factors led to an estimate of an around 2.3-fold difference between the cost per kWh of Superphénix and that of a PWR under economic conditions in 1982 [3].

This additional cost to construction was also noted during the construction of BN-800, which led Russia to temporarily freeze, mainly for economic reasons, the BN-1200 project, which is to take over.

Industrial domination of PWRs

PWRs have taken a dominant position in the market, which gives them advantages over competitors. These are essentially advantages of series effect and therefore of cost, as already described, as well as advantages in terms of industrial risk with products already in operation and better known.

This can be seen, for example, with boiling water reactors, which have many advantages, including lower operating pressure and a direct cycle to the turbine, but have great difficulty competing with PWRs.

As with any prototype, only strong political would allow the emergence of a new industry in place of a dominant industry. We see this currently with the small modular reactor NuScale being largely subsidized by American public funds, in an attempt to gain a foothold in the market.

Specific safety issues?

Comparing the safety of sodium-cooled fast reactors and PWRs is a bit like comparing a dog and a cat. Each of the channels has its advantages and disadvantages. You could even say that the advantages of one are the disadvantages of the other. At the time of Superphénix, the safety authority had concluded after specific requests and final analysis that the safety levels were equivalent (see Chapter 5 and Refs. [3, 6]).

Sodium-cooled fast reactors have many characteristics that enhance their safety, including:

- The absence of pressurization of the primary circuit and secondary circuits.
- The ability to evacuate the residual power from the reactor core by natural convection, without external water supply and with air always available as a cold source.
- The high margin between the temperature of sodium in normal operation and its boiling point.
- The excellent conductivity of sodium, which is favorable for the cladding and fuel temperatures.
- The favorable nature of the concept with regard to radiation protection.
- The significant thermal inertia of the primary circuit, which provides significant grace periods before putting safety systems into service.
- The simplicity of the conduct of the core and the absence of neutron poisons (no poisoning effect such as the xenon effect, unlike thermal spectrum reactors).
- Efficient trapping by sodium of the main fission products (in particular iodine and cesium).

On the other hand, a number of sensitive points are well known, including:

- The core is not in its most responsive configuration.
- A large part of the core can have a positive void effect.
- The power density is generally high.

- Sodium reacts chemically with many elements, in particular with water, air, and concrete, resulting in energy releases that can be significant, as well as the production of hydrogen during the reaction with water. On contact with air, aerosols from a sodium fire turn into soda, then sodium carbonate, before quick conversion to completely harmless sodium bicarbonate.
- Sodium reacts with uranium-plutonium oxide (when this type of fuel is used), which has led to the fuel being placed inside sealed ducts and the development of a very efficient system for detecting possible gas leaks.
- This same reactivity leads to potential sodium-water reactions at the level of the steam generators.
- The opacity and temperature of liquid sodium make it difficult to inspect structures under sodium.
- Although for certain components provisions can be made in the design to facilitate interventions and replacements, these remain difficult for circuits and sodium components.
- The duration of the unloading of the core assemblies is longer than in a water reactor. The unloading of fuel elements other than sodium can be done only after a long cooling time.

All these more delicate points are taken into account in the safety analyses and in the design of the reactor and make it possible to ensure a level of safety equivalent to or greater than that of water reactors. The significant operating experience feedback from these reactors [3,4] has enabled their design to be improved.

We can give three examples here:

- For the positive void effect, the work carried out within the framework of the French ASTRID project made it possible to design a core with zero or very low void effect. It is this core design that was taken up in the heart of the ESFR SMART project described in Chapter 5.
- For sodium fires, we will see in Chapter 5 the measures taken in the ESFR SMART project to confine the primary sodium and secondary sodium by reduction of piping length and containment by double wall. These provisions are inspired by the work done at Superphénix to prevent these sodium fires. Their detection and mitigation have been proven (see chapter 14 “Sodium leaks and fires” in Ref. [2]).
- At the level of the steam generators, the PFR and BN-350 reactors had significant sodium-water reactions due in particular to the ignorance of certain phenomena, such as that a high-pressure steam jet entering the sodium, as a result of a leak, acts like a blowtorch and causes new leaks (phenomenon called “wastage”). These phenomena are now well known. We know how to limit their occurrence and quickly detect and control sodium-water reactions. By way of example, see Chapter 9 of Ref. [2] for a description of the five sodium-water reactions that occurred and were controlled in Phénix.

It is not specific safety problems that will hamper the possible development of sodium reactors in the future, given the very good knowledge of this concept and the associated risks, which has been acquired over the past years.

However, the spectacular nature of the potential chemical reactions of sodium with water has had an unfavorable impact on public opinion, which is heavily exploited by opponents of nuclear power in general and of fast reactors in particular.

The proliferation risk

Much attention has been focused on the fuel cycle, particularly the extraction of plutonium from fuel through reprocessing extract, making it possible to manufacture mixed oxide (MOX) fuel based on natural or depleted plutonium and uranium.

Under the Carter administration in the United States, proliferation considerations demonized plutonium, which almost became a dirty word in the United States. This led to the cessation of reprocessing of American fuels in 1977. This fear was based on the idea that a chemical separation of plutonium from other compounds found at the end of irradiation of a fuel or of a fertile assembly is easier to implement than isotopic separation processes for ^{235}U, which is found in natural uranium. The shutdown was also favored by a lower financial cost of operation in the short term since no industrial-scale reprocessing of the spent fuel needed to be done. This position was confirmed in 1993 by the Clinton administration. As a result of this abandonment, the initial fissile component of the fast fuel was no longer available and the value of the fast cycle disappeared. You can run a fast reactor without plutonium, using enriched uranium as a fissile. This is what the Russians have been doing for a long time in BN-350 and BN-600. But we are clearly losing the advantage of the fast reactors, which is to operate with waste and not to use uranium-235. In these conditions, the United States closed their Fast Flux Test Facility reactor in 1993, and virtually ended their program on fast production reactors (although funds were recently allocated for the construction of a fast sodium research reactor). It can be noted, however, that unlike the risks associated with uranium enrichment, those associated with plutonium extraction have not been the subject of an international treaty such as the Non-Proliferation Treaty.

It should be remembered, however, that the qualities of plutonium required are not the same for civilian use in the production of electricity and for military use. The different Pu isotopes have different behaviors, which are not a problem for a fast reactor (whereas for a PWR they limit the amount of Pu recycling in the reactor). But these isotopes can become prohibitive for military use [7]. In the thermal spectrum, ^{240}Pu and ^{242}Pu crack little under the impact of neutrons, and ^{240}Pu, which is a very strong absorber of these neutrons, therefore constitutes a neutron poison. It is considered that, beyond a rate of approximately 7% of these even isotopes, plutonium can no longer be utilized for military use. In addition, ^{242}Pu has a high rate of spontaneous fission, which can trigger a reaction premature. ^{241}Pu poses another problem for the storage and handling of weapons, because over its 14-year life span it produces americium-241, the radiation of which will become increasingly troublesome. Finally, ^{238}Pu decays 240 times faster than ^{239}Pu and therefore releases much more thermal energy. Its prohibitive content seems to be around 2%. In conclusion, military use of plutonium requires ^{239}Pu.

When uranium-238 absorbs neutrons, it will first produce ^{239}Pu, then, but mainly in a thermal reactor, the isotopes 240, 241, and 242. We can see that the arrangement of outer covers of natural or depleted uranium on sodium-cooled fast reactors can be a source of ^{239}Pu production. It is also necessary to be able to manage and reprocess these blankets. In practice, the nuclear states had built dedicated factories to produce this plutonium, which were closed after the end of the Cold War and the corresponding agreements. Russia is gradually shutting down its dedicated RBMK reactors. The United States closed 14 reactors producing plutonium, and France closed the G1, G2, and G3 reactors located at Marcoule. Agreements to reduce nuclear stockpiles have even considered the use of military-grade plutonium in civilian reactors, notably BN-600 in Russia.

It should still be noted that this demonization of Pu in terms of proliferation, demonization linked to the original sin of Nagasaki, no longer makes much sense today. Recall that the two bombs launched on Japan were the Fat Man plutonium bomb on Nagasaki (about 6 kg) and the Little Boy bomb with enriched uranium on Hiroshima (64 kg, of which only 700 g participated in the fission).

A bomb with ^{233}U was also tested later (Fig. 4).

In terms of proliferation, a country that wants to acquire atomic weapons will clearly prefer to buy uranium, which is widely available on the market, or extract it from soil and enrich it by ultracentrifugation techniques, which are become relatively simple. The Pu route turns out to be much more difficult. It requires a reprocessing plant, associated with reactors in operation. In addition, the evolution of plutonium isotopes requires, for military plutonium, complex periodic operations to "clean" the americium, or to replace it with "new" Pu.

FIG. 4

View of the Little Boy enriched uranium bomb.

All countries that have sought to acquire nuclear weapons, including, recently, Iran or, less recently, South Africa, have simply bought the equipment to make centrifuges, to produce enriched uranium.

However, this demonization of Pu remains an obstacle to the development of fast reactors in some countries today. In France, this fear, in the context of the Cold War, contributed to provoking in the 1970s an opposition to the Superphénix project, which was much more important and aggressive than the vis-à-vis construction of a fleet of 58 PWR reactors at the same time.

Existence of reprocessing plants

Building a fast reactor where the fissile is enriched uranium would be of little benefit. Fuel reprocessing possibilities are needed to exploit the advantages of fast reactors operating without ^{235}U and with the byproducts of the use of ^{235}U in water reactors: depleted uranium upstream and plutonium in water reactors downstream.

However, these reprocessing plants are expensive and complex.

Today, France systematically reprocesses all its fuels and has the capacity to extract plutonium on an industrial scale (Fig. 5).

FIG. 5

The French reprocessing plant at La Hague.

Russia also has reprocessing plants. Japan began construction of a factory in Rokkasho in 2012, but it is not yet operational. China and India are in the planning phase. No other countries are developing such reprocessing plants.

This greatly limits the number of countries that could be attracted by the development of this type of reactor.

Fast reactor fuel manufacturing

Plutonium and uranium for fast reactors can be used in several forms, including oxide (as for water reactors), metal, carbide, and nitride (see Chapter 5).

Oxide fuels are those for which industrial experience feedback is the most important, in terms of manufacturing, irradiation, and reprocessing. This type of fuel is described below.

The fabrication of sodium-cooled fast reactor fuel is technically more difficult than that of water reactors, with successive operations of co-grinding, compression, sintering, and rectification.

You will find a description of it in Chapter 7 in Ref. [3], on Superphénix (Fig. 6).

These operations had been carried out for Superphénix in glove boxes with "clean" plutonium. Manufacturing tests for a small number of pellets were resumed as part of the ASTRID project on the MELOX facility located in Marcoule, dedicated to the manufacture of MOX fuel for water reactors.

But France does not currently have dedicated fast MOX fuel fabrication lines, since the closure and dismantling of the workshop that produced these fuels for Phénix and Superphénix, in Cadarache.

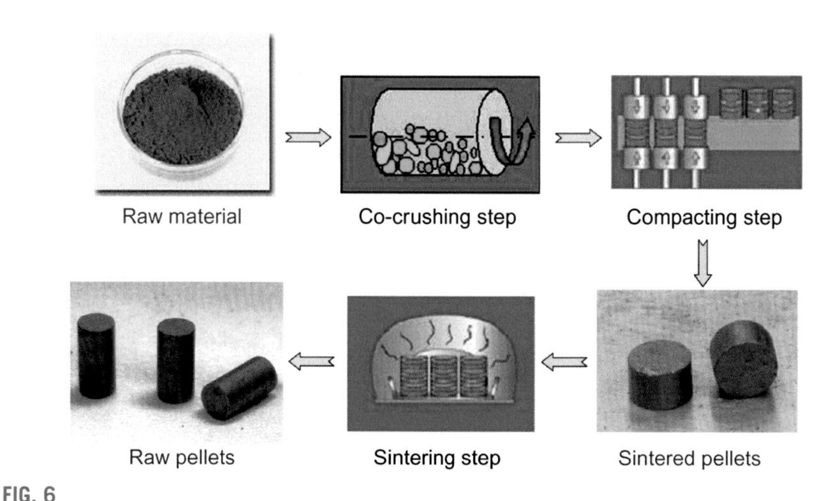

FIG. 6

Manufacturing steps of the pellets of a sodium-cooled fast reactor MOX fuel.

Russia has long operated the BN-600 reactor on enriched uranium fuel. With the startup of BN-800, work on controlling MOX fuel has made significant progress toward the stated objective of a complete shift to MOX fuel in BN-800 in the short term. This is in the logic of the search for a closure of the cycle, as expressed in the framework of the comprehensive Russian project "Proryv."

It should be remembered that, according to the Russian-American military plutonium management and disposal agreement, signed in 2000, the Plutonium Management and Disposition Agreement (PMDA), each of the countries must recycle around 34 tons of their respective plutonium stocks. For this, Russia has developed two tools: the BN-800 reactor and a MOX manufacturing plant in Krasnoyarsk, GKhK. With these two tools, Russia can use this plutonium for civilian purposes while respecting these agreements.

The next planned steps include:

- Partial then full loading of BN-800 with this MOX fuel manufactured at GKhK
- Fuel reprocessing
- Industrial demonstration of the closed cycle specific to fast reactors BN800 fuel

The fuel fabrication plant developed in Krasnoyarsk was built as part of the Russian Proryv project, which aims to develop cycle closure technologies and is based on two types of fast reactors not yet fully separated in Russia: sodium-cooled fast reactor and lead-cooled fast reactor (Brest).

This plant was built in the rock tunnels of a mine in a record time of 5 years. This MOX manufacturing plant, which supplies BN-800 fuels (Fig. 7), was inaugurated in

FIG. 7

View of the Russian sodium-cooled fast reactor BN-800.

2015. The first batch of fuels for BN-800 was supplied in 2017, and the announced production capacity is 400 assemblies per year. This production is carried out remotely with visualization by cameras.

The manufacture of this type of dedicated plant is therefore possible but constitutes an additional cost for the sector.

Social acceptance

For some people opposed to nuclear energy, the existence of a reactor that could operate without enriched uranium and burning of radioactive waste was a red flag, as it ensured the sustainability of nuclear power for thousands of years to come.

Indeed, the assets of the fast reactors constitute a response to many criticisms of nuclear power by these pseudo-ecologists, for example, on the future shortage of uranium or on waste management. This weakens their criticism of the nuclear energy they want to get out of, despite having no impact on global warming.

In France, the Superphénix reactor, during its 10 years of operation, was a constant target of environmentalists. It was stopped for 54 months for administrative reasons, that is, it was in working order but without authorization to operate. Following an election, the antinuclear forces finally succeeded in implementing its closure after a year of perfect functioning (for more details, see Chapter 4 of Ref. [3]).

Note that the above example represents a certain political period in France. This political opposition varies greatly from country to country. It can even change within a country depending on media and political circumstances, and awareness of the advantages of nuclear power over other sources of energy.

However, sodium and plutonium are products demonized by their opponents. Plutonium carries Nagasaki's original sin, and it is easy to make movies where sodium explodes in water. These elements do not facilitate social acceptance of the sodium-cooled fast reactors.

The loss of skills

It was difficult for nuclear sectors to operate in perpetual zigzag, following the laws of the market, regulatory changes, and political fluctuations. The time involved in development, construction, uptime, and dismantling of a reactor is generally on the order of a century.

We have recently seen these difficulties with the construction setbacks of the French European Pressurized Reactor (EPR) compared with the success of the Chinese EPRs, or with Westinghouse's setbacks for its AP-1000.

The same phenomenon, arguably at a greater scale, will occur when a fast reactor sector is restarted, where feedback has been forgotten by regulatory actors, operational or industrial, for decades.

Uranium availability

In the 1950s, uranium was considered rare and difficult to extract. Breeder reactors were proposed as a solution to the problem of this scarcity because, by converting uranium-238 into plutonium, they can potentially increase the amount of fission energy that can be extracted from 1 kg of uranium by 100-fold.

Uranium is the chemical element with atomic number 92, symbol U. It is a member of the actinide family. Uranium is the 48th most abundant natural element in the Earth's crust. Its abundance is comparable to that of molybdenum or arsenic, but four times lower than that of thorium. It is found in trace amounts everywhere, including in seawater, which on average contains about three parts of uranium per billion water molecules. That is, about 4.5 billion tons of uranium are dissolved in the world's oceans, about a thousand times more than known terrestrial reserves.

The problem with the availability of uranium is therefore not its scarcity but the fact that it is often found in trace amounts, unlike iron ores, for example, which are highly concentrated in certain places. Its average concentration in rocks is 3 g/ton. Thus, a typical garden contains several kilograms of uranium. It is these low concentrations that make extraction difficult and expensive.

However, this natural uranium content in the rocks of the subsoil can vary greatly from one area to another, and research has led to the discovery of high-concentration deposits. For rocks to be classified as "ores," it is generally agreed that they must have a uranium concentration greater than 0.1%, or 1 kg of uranium per ton of rock. This is the minimum rate for operation to be generally profitable in today's market. Some deposits were discovered that had high-grade ores that could incorporate up to 20% natural uranium, such as in the McArthur River and Cigar Lake mines in central Canada.

In 2007, the uranium requirements of the world fleet of nuclear power reactors were 67,000 metric tons—approximately 180 tons/GW of annual production. These requirements have changed little. The International Atomic Energy Agency (IAEA) predicts that global nuclear capacity will increase and uranium requirements will increase between now and 2050, but the scenarios remain highly variable.

Every 2 years, the OECD Nuclear Energy Agency publishes a report "Uranium: Resources, Production and Demand"—also known as the "Red Book." This report takes stock of known resources and consumption scenarios. The 2020 report [8] shows that the uranium market has remained depressed. Production fell slightly from 2016 (63,000 tU) to 2018 (53,516 tU), recovering slightly in 2019 (54,224 tU). This condition has resulted in low uranium prices. This has led to the closure of some mines (pending a hypothetical recovery) and a significant drop in prospecting. Global exploration spending reported in 2018 dropped by 75% from 2012.

In the long term, on a global scale, the amount of low-cost recoverable uranium will almost certainly well above the numbers reported in the Red Book. If plausible estimates of geological abundance are used, the quantity of uranium still to be discovered at recovery costs of up to \$130 (USD)/kg remains very large if exploration resumes.

The price of uranium on the spot market rose significantly above $130/kg in the late 1970s, and again after 2005. Apart from these two periods of imbalance between supply and demand, prices were less than $50/kg. The price spike in the 1970s was due to the expectation of a huge expansion in nuclear capacity. This expectation was not fulfilled, but large stocks of uranium had been built up and were sold over the following decades, leading to the closure of many uranium mines. Russia's sale to the United States of 500 tons of military-grade uranium from surplus Cold War weapons, for use after dilution in civilian reactors, also fueled half of the United States' nuclear capacity. This prolonged the period of low uranium demand.

The last uranium price spike in 2007, at $130/kg, reflected, at least in part, the expectation, compounded by speculation, that there could be uranium shortages due to the capacity of uranium mining not returning to the level required to support the then growing demand (Fig. 8).

In any case, unlike the situation with oil or gas power plants, the cost of uranium fuel can double without having a significant impact on the cost per kWh of nuclear electricity. Even at $130/kg, the cost of uranium is only about 5% of the cost of electricity produced by a light water reactor.

However, if nuclear power were to increase its capacity to replace fossil fuels, uranium consumption could increase significantly. For example, China, with its 51 reactors in place in 2021, produces only 5% of its electricity from nuclear power. However, even with a high level of nuclear reactor construction, uranium would remain relatively abundant, and with bearable price increases, in this century. On the

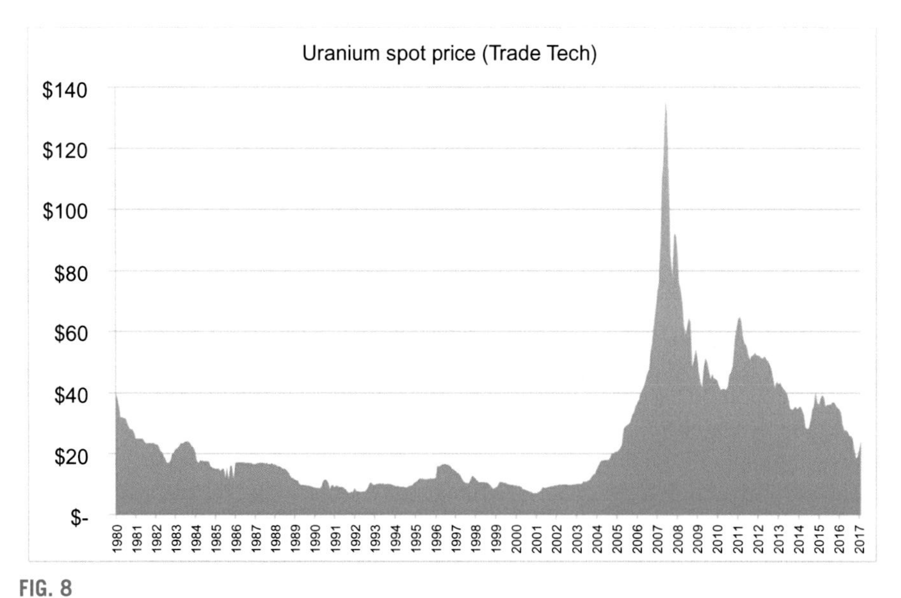

FIG. 8

Trend in uranium spot prices.

other hand, if we consider scenarios where the growth of the nuclear fleet is necessary to replace fossils, we encounter strong uncertainties and tensions in the market.

This market is very speculative, and price increases could appear in the event of a lag between increased demand and resumption of prospecting. For example, the price of crude uranium, known as yellowcake, reached its highest level in 2021 since 2014, thanks to a new investment trust led by Canadian asset manager Sprott. Investors are betting that nuclear power will play a key role in the move away from fossil fuels and that a lack of new uranium mines will mean the price must rise. The Sprott Physical Uranium Trust has recovered approximately 6 million pounds of physical uranium, valued at approximately $240 million, since its launch on July 19, 2021, helping to drive uranium prices to over $40 a pound, compared with $30 at the beginning of the year. Global mine supplies are expected to be around £125 million in 2021. Its aggressive purchase will put pressure on operators, who need to secure the product's supply for power generation. China projects a surge in nuclear capacity over the next decade. In addition to the holdings of an acquired fund, Sprott currently holds 24 million pounds of uranium, valued at approximately $1 billion, in the form of yellowcake. Other financial players also bought the commodity on the bet that its price would increase. Yellow Cake Plc, listed in London in 2018, holds approximately 16 million pounds of uranium. "This has been a key factor in the 30% increase in the price of the metal in 2021," said Nick Lawson, managing director of brokerage firm Ocean Wall. Demand for uranium is expected to rise from around 162 million pounds this year to 206 million pounds in 2030—and even more to 292 million pounds in 2040—according to the World Nuclear Association, largely thanks to the increased power generation in China as Beijing seeks to reduce greenhouse gas emissions. At the same time, the uranium supply is expected to fall 15% by 2025 and 50% by 2030 owing to lack of investment in new mines. Financial players are clearly accelerating the rise in prices, but this would not happen if there was no vision of a fundamental and substantial deficit ahead, some analysts have said.

It is clear that a fleet of fast reactors operating for thousands of years with stocks of depleted uranium already available represents a much more controlled situation. On the other hand, the argument of the need for fast reactors to avoid a future scarcity of uranium does not appear fully relevant for this century.

Conclusions

After reading this chapter, the reader may be discouraged about the future of these sodium-cooled fast reactors. However, a few nuances must be added.

This fast reactor process requires a very significant effort, of scientific and technical investment for the country undertaking it. Moreover, the benefits of this heavy investment cannot be seen until decades later. We are therefore not expecting short-term benefits, but rather counting on the benefits for the community in the coming century. Moreover, it is this same short-term thinking that is in the United States now leading to the closure of certain nuclear power stations, temporarily unprofitable compared with the provisionally low prices in the fossil fuel market.

Relying only on PWRs/open cycle to replace fossil fuels in the future could lead to speculation and tensions in the uranium market. Nevertheless, the availability of uranium at a bearable price seems assured for the following decades.

Note also that the open cycle has allowed significant financial benefits for PWRs in the countries concerned, because reprocessing is much more expensive than interim storage in swimming pools. However, it is clear that this is an unsustainable solution in the long term, compared with the virtuous cycle of using natural resources and managing waste from fast reactors.

Regarding nonproliferation, controlled reprocessing allows real and centralized management of plutonium as well as its use as fuel. Long-term storage of tons of plutonium in spent fuel not intended for this purpose indeed seems to be a way of postponing the problem for future generations. Moreover, the facts show that proliferation is much riskier with enriched uranium than with plutonium.

It is not the operation or the safety of these reactors that is the problem; it is rather their significant additional cost and the corresponding political commitment necessary, for real ecological benefits in the long term, after a long transitional period of nonprofitability, compared with the free energy market based on fossil fuels and current nuclear power. The social acceptance of this action is necessary.

This period between product launch and possible commercial takeoff can last tens of years for these sodium-cooled fast reactors. In Fig. 9 we show the valley of death and where many projects that are not immediately profitable are bogged down in this period. Sodium-cooled fast reactors, in particular, can be bogged down because of the high investments required with great continuity in the reactor/reprocessing plant/fuel fabrication plant triptych, and the long duration of this period of nonprofitability and risk.

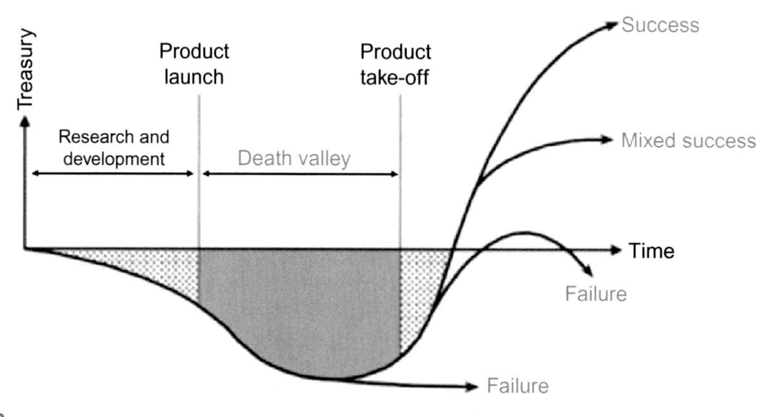

FIG. 9

Valley of death for long-term costly projects with no immediate profitability.

References

[1] https://world-nuclear.org/information-library/current-and-future-generation/fast-neutron-reactors.aspx.
[2] J. Guidez, PHENIX The Experience Feedback, Editor EDP Sciences, 2013.
[3] J. Guidez, G. Prele SUPERPHENIX Technical and Scientifical Achievements Editor Springer, 2016.
[4] J. Guidez, Coll, Lessons learned from sodium cooled fast reactor operation and their ramifications for future reactors with respect to enhanced safety and reliability, Nucl. Technol. 164 (2008) 13.
[5] J. Guidez, Review of the experience with worldwide sodium fast reactors operation and application for future reactor design, in: IAEA International conference on research reactors November 2007 Sydney/Australia, 2007.
[6] Report No. 1018 of the parliamentary commission of inquiry into Superphénix and the fast neutron reactor sector, recorded on June 25, 1998 and published in the official journal on June 26, 1998.
[7] H. Metivier, Plutonium Mythes et réalités, EDP Sciences, 2009.
[8] AEN, IAEA report, Uranium Resources, Production and Demand, Redbook, 2020.

CHAPTER

What fuel for a sodium-cooled fast reactor? 5

Abstract

A sodium-cooled fast reactor uses a fuel composed of a fissile element (generally plutonium) and a fertile element (depleted uranium or natural uranium) in proportions of approximately 20% and 80%, respectively. These proportions vary according to the origin and quality of the Pu used.

This fuel, called mixed oxide (MOX), has been tested and studied in four different forms: oxides, metal, carbide, and nitride.

We outline the qualities and defects of these four forms of fuel both during reactor operation and on completion of the fuel cycle, i.e., in reprocessing and fabrication.

It appears that the choice of fuel is linked to historical reasons, and to different utilization strategies depending on the country.

In conclusion, the four types of fuel remain potentially acceptable candidates as fuel for a fast reactor with sodium coolant even if they vary in terms of experience feedback and level of qualification or maturity.

According to the current state of knowledge, a pragmatic choice would be oxide, which has the greatest operating experience feedback over the entire cycle: reactor operation, manufacturing, and reprocessing.

Keywords

MOX, Carbide, Nitride, Oxide, Fabrication, Fuel, Reprocessing

Fast Reactors. https://doi.org/10.1016/B978-0-12-821946-1.00005-X

Introduction

Four types of fuels have historically been tested and used for sodium-cooled fast neutron reactors.

The first American reactor (CP1), developed at the end of 1942, used as fuel a mixture of uranium oxide and metallic uranium. However, in 1951 the first fast neutron reactor (Experimental Breeder Reactor-II (EBR 2)) used fuel based on enriched uranium metal. Thereafter, American research nuclear reactor technology remained faithful to these enriched uranium and uranium-plutonium fuels in the form of an alloy with Zr in EBR 2 and in all their other achievements and projects: Fast Flux Test Facility (FFTF), Power Reactor Innovative Small Module (PRISM), Versatile Test Reactor (VTR), etc.

Although metallic fuel was the first fuel used for fast neutron reactors, it was supplanted by oxide in most of the fast neutron reactors that were developed around the world, particularly in France. During the 1960s, the French Alternative Energies and Atomic Energy Commission (CEA) first worked on a metal alloy, UMo, before finally switching to mixed uranium and plutonium oxide fuel for the RAPSODIE reactor. This orientation was also linked to the choice of reprocessing based on the PUREX hydrometallurgical process developed for graphite gas and light water reactor fuels, a choice that was better suited to U-Al and oxide alloy fuels. This oxide fuel was then used at Phénix and Superphénix. It is the only fuel that has industrial experience in reprocessing and reuse in the reactor, thus making it possible to complete the fuel cycle (see a description of this industrial experience in Appendix 2).

Mixed carbide fuels of U and Pu also have specific advantages that have led to their in-depth study for their use in SFRs in the 1960s and 1970s in France but also in the United States and at the Institute of Transuranium Elements in Karlsruhe, Germany. Later, in the early 2000s, they were also studied within the framework of certain reactor concepts, such as fast reactors cooled with a gas coolant. Over the past 20 years, it was mainly India that manufactured, tested in their small reactor (Fast Breeder Test Reactor, FBTR), and reprocessed mixed carbides, which were sometimes very rich in plutonium (70%). Today, all countries, including India, have abandoned this fuel in their new projects. Only FBTR, still in operation, continues to use UPuC fuel with a high plutonium content.

Nitride fuels, quite similar to carbides, have been studied in several frameworks but to a lesser extent than carbides. Recently, they have regained interest, but specifically in Russia for lead fast reactors, where their high density makes it possible to compensate for the spacing of the pins imposed by the low speed of lead, caused by corrosion problems. Within the framework of the BREST project (see Chapter 3), nitride fuels have been manufactured and are being tested on BN-800. In all other countries, it has been abandoned.

In conclusion, today there are four types of fuel available for sodium-cooled fast reactors (SFRs) (Table 1).

The purpose of this chapter is to outline the advantages/disadvantages of these four types of fuel. This analysis includes, in addition to a comparison of these fuels in operation in a reactor, a comparison of the facilities for reprocessing and fabrication.

This analysis is restricted to fuels containing uranium and plutonium. In fact, fuels based on enriched uranium alone are only useful in the initial phases of developing SFR technology. The final objective of fast neutron reactors is to complete the fuel cycle by making best use of the available fertile material and by recycling the plutonium.

The analysis of thorium fuels was also excluded. Some samples have been tested in the reactor and in fuel pins, but thus far no fast reactor has operated with thorium-based fertile material, mainly because of the abundance of depleted uranium available as fertile material. Only India has tested a thorium-oxide blanket in the experimental FBTR fast breeder reactor, to explore the possibilities of reprocessing this type of fuel.

Table 1 Main characteristics of fuels for fast neutron reactors (IAEA Nuclear Energy Series No. NF-T-4.1 (2011) "Status and Trends of Nuclear Fuels Technology for Sodium Cooled Fast Reactors").

Properties	$(U_{0.8}Pu_{0.2})O_2$	$(U_{0.8}Pu_{0.2})C$	$(U_{0.8}Pu_{0.2})N$	U-19Pu-10Zr
Theoretical density (g/cm^3)	11.04	13.58	14.32	15.73
Melting point (K)	3083	2750	3070	1400
Thermal conductivity ($Wm^{-1}K^{-1}$) at 1000–2000 K	2.6–2.4	18.8–21.2	15.8–20.1	40–40
Crystal structure	Fluoride	NaCl	NaCl	Alfa
Breeding ratio	1.1–1.15	1.2–1.25	1.2–1.25	1.35–1.4
Swelling	Moderate	High	Moderate	High
Handling	Easy	Pyrophoric	Inert	Inert
Compatibility: clad	Average	Carburization	Good	Eutectics
Compatibility: coolant	Average	Good	Good	Good
Dissolution and reprocessing	Good	Demonstrated	Risk of C14	Amenable for pyro-reprocessing
Fabrication/irradiation experience	Large and good	Limited	Very little	Limited

Oxide fuel

Oxide fuel currently has the greatest amount of feedback experience available.

Uranium and plutonium oxide fuel was chosen by the French sector, mainly because of its compatibility with the PUREX-type hydrometallurgical reprocessing process developed in France, and is used today for the reprocessing of pressurized water reactor fuels.

MOX fuels $(U,Pu)O_2$ with plutonium content above 15% have the main disadvantage of reacting with sodium to form sodium uranoplutonate $Na_3(U,Pu)O_4$, a compound whose density and thermal conductivity is respectively twice and three times lower than that of oxide. The use of a sodium seal is therefore prohibited with MOX fuels. A gas seal, generally helium, is preferred for the oxide fuel element because it is the best conductive neutral gas.

This helium seal has advantages and disadvantages.

The advantages of the helium-sealed oxide fuel pin include:

- Easier manufacture of oxide pins compared with sodium joint pins.
- Possibility of recovering the fission gases released in the pin in a free volume placed in the lower part of the pin, where the temperature in normal operation is that of the sodium inlet temperature (~400°C).
- Risk of clad-fuel chemical interaction and clad corrosion, a priori lower because the fission products are not conveyed by a liquid metal seal.
- Reprocessing of the pins similar to that of the water reactor rods, not requiring prior operations to eliminate the sodium in the pins; these operations would otherwise be very complex.

The disadvantages of a gas seal include:

- Regarding the heat of the fuel, the gas seal, even if its size is reduced, with a manufacturing thickness on the order of 0.1 mm, will lead to a sharp increase in temperature at the start of its use during the first powerup. As the conductivity of the MOX is lower than that of the metal alloy, at identical linear power, this leads to a core temperature at the start of use of around 2400°C compared with 1000°C for a sodium joint metal.
- At high burnup, the significant release of fission gases from the fuel leads to an internal pressurization of the pin of a few tens of bars, which can be significant in certain incidental operating situations because it is likely to generate mechanical stresses that can damage the clad.

The advantages of MOX fuel include:

- Its melting temperature is very high; hence, there is a relative inertia with respect to the clad steel.
- The operation of the MOX at high temperature in an SFR leads the oxide to deform by creep. This makes it possible to avoid, in most incidental operating situations, a mechanical oxide clad interaction liable to damage the clad.

- Its industrial manufacture (more than 100 tons of MOX manufactured, more than 400,000 fuel pins produced in France alone), its use in power reactors in various countries around the world, and its reprocessing on a large semiindustrial scale, particularly in France, places it at a level of qualification without equivalent among all the other potential fuels for the rapid sector.

However, oxide fuel has a number of intrinsic disadvantages related to its physical properties:

- Because of its lower density, it is more difficult to obtain long-lived cores than with metallic fuel. However, burnup rates close to and above 20% have been achieved on experimental fuel elements respectively irradiated in Phénix and Rapsodie, because the thermomechanical and physicochemical behavior of the MOX has no intrinsic limit.
- It has a lower expansion coefficient, which leads to poorer feedback.
- Its lower thermal conductivity leads to significant axial thermal gradients and may lead to its exclusion from certain uses (such as gas fast reactors).
- Its chemical reactivity with sodium during clad rupture requires the use of detection systems in reactors: DRG (cladding rupture detection) and DND (delayed neutron detection). Beyond a DND threshold, an emergency shutdown is automatically triggered before fissile material can be significantly disseminated in the primary circuit. The data acquired in the Phénix reactor, and before in Rapsodie, as well as the numerous experiments carried out on broken pins in the sodium loop of the SILOE reactor generated significant feedback on the subject (there was no cladding failure in Superphénix).
- Several concomitant factors make its general behavior in accidental transients worse than with metallic fuel because the higher fuel temperature decreases the margin on melting by the same amount, there is an exothermic reaction with the sodium in the event of cladding failure, and the lower axial expansion coefficient compared with that of the metal alloy leads to weaker counterreactions.
- The feedback Doppler effect is weaker for the oxide fuel than for all the other fuels (carbide and nitride included) during fast reactivity transients owing to a smaller difference between the operating temperature and the temperature of fusion.

In general, the advantage of the high melting point of the oxides is partially canceled out by high fuel temperature under the required linear power operating conditions, where the radial temperature gradient in the pellet can reach 5000°C/cm. To increase this margin on fusion, pellets with a central hole were utilized in Superphénix. This avoids any risk of core meltdown during normal operation (Fig. 1).

For this annular fuel, reactivity insertion tests were also carried out on the CABRI experimental reactor to examine its behavior in the case of an accident. These tests (Fig. 2) showed fuel melting, but without cladding rupture thanks to the central holes in the pellets.

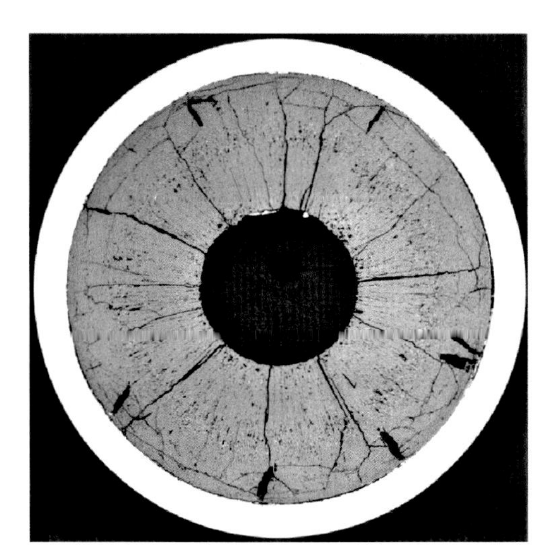

FIG. 1

Cross-sectional macrography of a Superphénix-type annular fuel after irradiation.

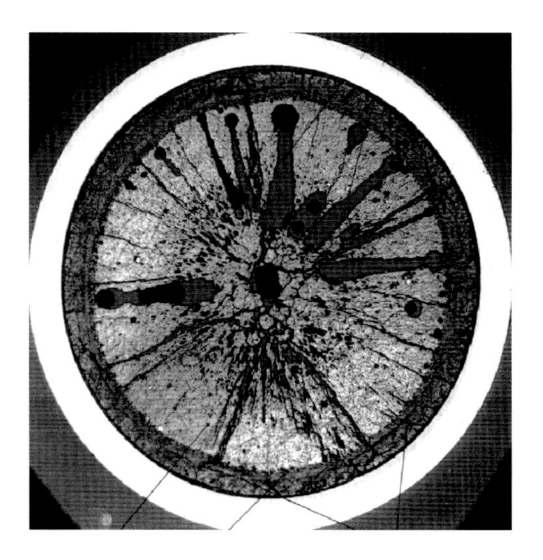

FIG. 2

Extensive melting of the fuel (light zone) but without cladding failure after rapid reactivity insertion test.

This design has been validated on Superphénix.

Dispersion of oxides in a metal (cermet) has also been studied to improve conductivity but without very convincing results.

At the international level, the significant feedback experience for this fuel, in the reactor and in the case of an incident or accident, as well as at the reprocessing level, make it the reference fuel in fast reactors currently operational in Russia (BOR-60,

BN-600, BN-800), India (PFBR, to come), China (CEFR and the CFR-600 MWth, to come), and Japan (Joyo, Monju), The countries that have chosen to close the cycle have selected oxide fuel because it is the only combustible material to have demonstrated semi-industrial feasibility of the process over its life cycle: in the reactor and upstream and downstream of the cycle (Appendix 2).

Metallic fuel

Metallic fuel has a number of advantages:

- It has a high density, which makes it possible to envisage a more compact core and optimized regeneration gain (or doubling time: time required for the mass of Pu to double).
- The absence of oxygen atoms makes it possible to have a "faster" neutron spectrum and, consequently, a better economy of neutrons, hence a higher yield in regeneration and a longer residence time.
- The Doppler coefficient is lower owing to the greater hardness of the spectrum. However, this low Doppler is compensated by the low operating temperature as well as a high expansion coefficient, which allows excellent behavior in accidental transients.
- The UPu10Zr alloy has very good thermal conductivity, which leads to a maximum fuel operating temperature of less than 1000°C. Thus, despite a low melting temperature, the melting margin ($T_{fusion} - T_{nom}$) remains similar to the oxide.
- The expansion coefficient for UPuZr ($\sim 20 \times 10^{-6}\,K^{-1}$) is greater than that for the oxide ($(U,Pu)O_2$, $\sim 10 \times 10^{-6}\,K^{-1}$). It therefore has better behavior in thermal transients and severe accidents.
- The manufacture by casting of UPuZr alloy bars is, it seems, no more complex than the manufacture by powder metallurgy of fuel pellets. We retain from the experience feedback a simple manufacture (at least for a bar of 1 m) with stable behavior whatever the manufacturing defects, therefore a specification of the fuel that seems not very restrictive. Nevertheless, the fabrication of the fuel element with metallic fuel is more complicated than that of an oxide fuel pin because of the use of a sodium seal between the fuel rod and the clad. In fact, the very large radial clearance between the bar and the clad is imposed by the strong swelling of the UPuZr. This gap, if it were filled with gas, would lead to a very high thermal gradient; therefore, reaching the core melting temperature of the fuel would be inevitable. This requires the use of a sodium seal between the bar and the clad, to prevent the UPuZr from reacting with the Na.
- The achievement of high linear power, on the order of 1000 W/cm would be possible thanks to the good thermal conductivity if there were no limitation of the operating temperature to reduce the fuel-clad chemical interaction.
  In normal operation and during incidental transients, the creep capacity of the UPuZr alloy of the bar makes it possible to avoid any strong mechanical interaction with the clad.

- High combustion rates (> 15%) have already been achieved without difficulty, with an appropriate fissile material filling density.
- The addition of minor actinides is possible without leading to a different behavior of this fuel under irradiation (cf. Métaphix irradiations in Phénix).
- In power transient, the metal has a significant melting margin compared with the oxide thanks to a thermal conductivity almost 20 times greater at high temperature and a mechanical interaction with the clad reduced thanks to creep raised. The only existing power transient tests on preirradiated metal pins were carried out in the United States in the TREAT reactor (six tests).
- In transient loss of coolant, a metal fuel pin has very good performance in terms of restitution of stored energy linked to its low operating temperature compared with oxide fuel elements (combined effects of its very good thermal conductivity and the sodium seal).
- In severe accidents (general meltdown), the advantage of the metal fuel element stems from its low stored energy.

However, it has a number of drawbacks:

- Its low melting temperature and its high swelling require the use of a sodium seal between the fuel and the clad to limit the thermal gradient in a generally large clearance, allowing a filling density on the order of 75% DTh to be reached.
- Under the operating conditions of a fast reactor, its significant swelling under irradiation makes it necessary to limit the filling density between 70% and 75% DTh, which somewhat weakens the advantage of its high intrinsic density. It should also be noted that the elongation of the fuel column due to thermal expansion, but especially due to the swelling of the fuel, generates a loss of reactivity over time.
- During irradiation, UPuZr, in conjunction with certain fission products (lanthanides) can react chemically with the austenitic or ferritic steel clad and lead to the formation of clad-fuel eutectics, leading to the embrittlement of the clad. Such behavior requires one to limit both the fuel temperature by limiting the linear power and the cladding temperature (620°C instead of 700°C for oxide cores). Consequently, the core outlet temperature and the thermodynamic efficiency are reduced compared with an oxide core. To optimize performance, a remedy for this chemical reaction with the formation of a eutectic would be, for example, the use of an internal coating of the clad, which remains to be developed but which would complicate the manufacture.

To address these drawbacks, solutions have been proposed:

- To compensate for the intrinsic swelling, the thickness of the sodium seal must be optimized according to the target combustion rate. For combustion rate comparable to those obtained for oxide fuels, a filling density of approximately 70% DTh is targeted. We thus lose part of the advantage of fissile density of this fuel.

- Fast neutron reactor (FNR) operating conditions lead to a very significant release of fission gases on the same order as for the oxide. Consequently, the pin must be equipped with a large plenum that, owing to the presence of the sodium seal, must be positioned in its upper part, which excludes certain design options with plenum in the lower part.
- Recent studies suggest that the addition of 6% instead of 10% Zr would reduce the fuel-cladding chemical interaction but would reduce the melting temperature of the alloy by about 40°C, partially compensated by an increase in thermal conductivity.
- To limit the problems of eutectic formation with the cladding, an internal protective liner with an ad hoc metal (e.g., Zr, V) could be considered: this is the current and main area of research and development for this type of fuel. A buffer was also considered to replace the sodium joint.

At the international level, the United States and South Korea have chosen metallic fuel as the reference fuel for SFR. Other countries (India and China) consider it as the objective to be achieved after the oxide in particular to increase the regeneration rate. France has always ruled out the use of this type of fuel, which is not very compatible with its pyrometallurgical reprocessing processes.

In conclusion, the Zr-stabilized fuel of the U-Pu-10Zr type has been relatively proven in the United States up to the scale of a small core (EBR 2) and up to burnup rates equivalent to those obtained with the oxide. In addition, the tests carried out in TREAT have shown good behavior in extreme transients (Fig. 3).

FIG. 3

View of the site of the EBR-2 experimental reactor where metallic fuel was tested, as well as its fabrication and reprocessing, from 1965 to 1994.

The weak point of the metallic fuel element remains manufacturing and reprocessing with pyrometallurgical treatment, raising doubts about its large-scale industrialization. The application of a hydrometallurgical treatment of the PUREX type would require, to treat this type of fuel, additional process steps using chemical reagents that are often incompatible with the composition of the materials of the treatment devices currently used in the La Hague plants.

Carbide fuel

Carbide fuel has a number of intrinsic advantages:

- Better conductivity than oxide (about a factor of 10).
- A melting temperature on the same order of magnitude as that of the oxide.
- A high density, on the same order of magnitude as that of metal.
- No chemical reaction with sodium under nominal conditions.
- No eutectic reaction with the clad.
- From a neutronic point of view, the carbide is between oxide and metal, with a fast spectrum that is, however, close to that of metal.
- In power transients, carbide has significant margins with respect to the melting temperature, combining a very high conductivity (close to that of metal) and a very high melting temperature (close to that of oxide). On the other hand, the mechanical interaction with the cladding is the weak point of this fuel (significant expansion, gaseous swelling, and, above all, a low capacity to accommodate its own deformations by creep). The only existing power transient tests on preirradiated carbide pins were carried out in the United States in the TREAT reactor (10 tests on carbide).

This fuel has been tested via two concepts: gas seal (He) and sodium seal. Their advantages/disadvantages have already been mentioned for metallic fuel and oxide fuel.

A number of problems can be noted with the use of this fuel:

- It can cause significant embrittlement of the clad steel by carburization, which is much greater for the sodium seal concept.
- For the concept with helium seal, its swelling under irradiation is much greater than that encountered for oxide mainly in the case of operation at high temperature. This swelling is characterized by a strong gaseous component linked to a high stability of the intra- and intergranular bubbles. This significant swelling of the carbide is at the origin of a mechanical interaction with the clad even in steady state in normal operation. With a helium-sealed concept, it is mandatory to use low-density fuel (80%–85% DTh) whose manufacturing porosity remains as stable as possible under irradiation to be able to partly accommodate the swelling. The manufacture of such fuel is not easy.
- For the sodium seal concept, the carbide operates at relatively low temperatures owing to the very good thermal conductivity of the seal and the carbide. Under these conditions, the swelling of the carbide, solid and gaseous, remains limited

but still greater than what could be measured for an oxide. With this concept, it is possible to envisage a fuel of higher density, which, with sufficient thickness of the sodium joint, makes it possible to reduce the risk of severe mechanical interaction with the clad. Like metal, protection of the inner face of the clad must be considered to avoid excessive and damaging carburization.

- The carbide fuel is more difficult to manufacture than an oxide because of the use of pyrophoric materials, and a rather strict specification concerning its chemical composition. Indeed, the behavior under irradiation of the carbide is strongly dependent on its initial characteristics (contrary to the metal), content of impurities in particular oxygen, density, and microstructure (size of grains, proportion, nature and distribution of porosity). This imposes manufacturing specifications with tight tolerances. The challenge of making a good sodium seal is another factor to take into account, which does not facilitate its manufacture.

At the international level, the mixed carbide of U and Pu is a fuel that has been studied very intensively, in particular by the United States, Europe (result of the program from the 1970s to the 1990s with the NIMPHE and NILOC irradiations in the Phénix reactor), India, which utilizes a fully carbide core in their research nuclear reactor FBTR, and Japan.

It can be noted that India succeeded in sizing a carbide pin without a sodium seal; however, they achieved this by significantly reducing the linear power and, consequently, the temperature of the combustible, to limit swelling. The pin thus reached a very high burnup rate in the FBTR reactor (165 GWd/t). Notably, in recent years, India has succeeded in reprocessing two tons of irradiated carbide fuel.

Despite these few positive points, carbide fuel has now been abandoned all over the world (including in India) because of the difficulties of industrializing manufacturing, which would be much more expensive than that of oxide. Let us cite, for example, the obligation to put all the manufacturing workstations dedicated to the handling of powders in an inert and perfectly controlled atmosphere to reduce the risks of pyrophoricity. This point is also a brake on the reprocessing of this fuel by hydrometallurgical means. Another difficulty during reprocessing is the formation of carbonaceous species in solution, which can lead on the one hand to significant losses of Pu in the extraction raffinates, and on the other hand to unstable or even explosive phenomena (formation of red oils) related to the concentration of fission products.

Nitride fuel

Nitride fuel has virtually the same intrinsic advantages as carbide:

- Better conductivity than oxide (about a factor of 10)
- A high density on the same order of magnitude as that of metal
- No chemical reaction with sodium
- No formation of eutectic with the clad
- A strong Doppler effect of interest in certain safety studies

From a neutronic point of view, nitride is located, like carbide, between oxide and metal alloy, with a fast spectrum that approaches that of metal.

Like carbide, nitride can be used with sodium or gas sealed fuel element designs.

There are a number of problems specific to nitride in addition to those already mentioned for carbide:

- Unlike carbide, the melting temperature is not the property to be considered. It is the dissociation temperature that must be taken into account in the design studies. Indeed, at low nitrogen partial pressure, uranium and plutonium nitride dissociates at a much lower temperature level, around 1000 K below its theoretical melting temperature. This problem can lead to the formation of metallic phases rich in plutonium within the nitride fuel. For example, French and Japanese irradiations have shown the appearance of metallic Pu in the pellet/clad clearance, for irradiations carried out admittedly at fairly high linear power (700 W/cm for NIMPHE 2) but at relatively modest combustion rates. Furthermore, the risk of dissociation may increase in the event of cladding failure. Consequently, this problem must be taken into account in the design of nitride fuel pins with a helium seal.
- The other disadvantage of nitride is the production of ^{14}C by neutron capture on ^{14}N. To avoid this production, it would be necessary to carry out a very thorough preliminary enrichment (more than 99%) in ^{15}N, which is a feasible but expensive operation.
- There is a risk of embrittlement of the clad by nitriding in the presence of a Na seal.

The behavior under irradiation of helium-sealed nitride fuels was found to be less problematic than that of carbide under identical operating conditions. Its swelling under irradiation is a little less important than that of carbide at the same temperature. However, experience feedback remains limited worldwide, with about 10 times less experience accumulated over five decades for nitride than for carbide.

At the international level, nitride fuel has been studied particularly by Russia, but also by the United States, Europe, and Japan. Russia loaded their BR-10 mini-reactor with two enriched uranium nitride cores and is also considering a nitride feeder core for the BREST-300 lead-cooled reactor project. A fabrication line is under construction, and the qualification of this fuel is continuing in BOR-60 and BN-800.

Japan, after working on nitrides (JAERI) and metal (CRIEPI), finally returned to oxides more compatible with their PUREX reprocessing process used industrially in their Tokaï Mura and Rokkasho Mura plants.

Apart from Russia, which continues to carry out projects with nitride fuel, it has been abandoned by the other major nuclear countries, probably because of expensive industrial production requiring, among other things, very high enrichment in ^{15}N nitrogen and probably its recycling during reprocessing. Another negative aspect of nitride is its lack of chemical stability at high temperature, especially in the case of accidents (Fig. 4).

FIG. 4

View of the FNR BN-800 reactor, where a nitride fuel is currently being tested.

Fuel/manufacturing

Fuel elements with these four types of fuels can be designed in two ways:

- Sodium seal concept: mandatory for metal alloy, optional for carbide and nitride
- Concept with gas seal (He): mandatory for oxide, optional for carbide and nitride

The gas seal concept is much simpler to manufacture industrially than the sodium seal concept. For carbide and nitride fuels, the gas seal concept should therefore be favored, if possible.

The manufacture of annular sintered MOX oxide pellets remains difficult, but the specifications required at the end of manufacture are less restrictive than for MOX from water reactors. Feedback, particularly in France at the ATPu nuclear facility in Cadarache, showed a technique well mastered at the time (1970–99) as in the United States, Japan (Monju), and Germany (KNK II and SNR-300). More recently, Russia loaded the entire core of BN-800 with MOX oxide.

Metal is the only fuel that does not require the manufacture of pellets, only simple bars of metal obtained by fusion.

Nitrides require very thorough enrichment of nitrogen in ^{15}N.

The manufacture of carbides is delicate: the atmosphere must be perfectly controlled because of the risk of pyrophoricity of the carbide powders. Controlling the density of carbide pellets remains difficult; it is necessary to obtain after sintering a stable and high porosity (15%–20%) to accommodate the high swelling. Ultimately, very good control of impurities is essential, particularly oxygen, the residual content of which after sintering must not exceed a few hundred ppm.

In conclusion, there is a definite advantage to oxides and, to a lesser degree, to metal in terms of manufacturing and the corresponding industrial REX.

Fuel/reprocessing

Following the general logic of fast reactors, the fuel must be able to be reprocessed with a view to recycle the uranium and plutonium in the manufacture of the new fuel.

The metal fuel reacts with water, producing hydrogen with the sodium seal. The use of a hydrometallurgical reprocessing process based on dissolving the fuel in an aqueous solution is not impossible but more complex because of the reagents to be used (this has already been studied on UZr in the United States and France). In summary, considering the current state of knowledge, pilot-scale reprocessing of metal fuel by hydrometallurgical means is not impossible. However, large-scale industrialization of the process will require innovations in the choice of materials for the process equipment to be used (high risk of corrosion) and in the management of effluents, the containment of waste, and the recycling of reagents. Nevertheless, in all cases, to avoid the crystallization of zirconium fluoride, it will be necessary to limit the concentrations of uranium and plutonium at dissolution to approximately 50 g/L, resulting in a considerable drop in the reprocessing rates obtained for oxide fuel.

For metal fuel, the United States has chosen pyrochemical reprocessing. The fuel is dissolved in a chloride salt bath, and the uranium and plutonium are partially separated from the fission products by electrolysis with yields and decontamination factors that need to be improved. At this level, the process works but with separation and recovery efficiencies that remain much lower than those that can be achieved with the PUREX process for UOX and MOX fuels, for example. Therefore, U and Pu losses occur in the salts. Feedback on the application of this process remains weak today and is mainly limited to the tests carried out on feeder fuels EBR 2 and FFTF at scales closer to what is expected from a pilot project than an industrial chain (several hundred kilograms) (Fig. 5).

The best feedback from FNR fuel reprocessing is at the level of oxides with the PUREX process. After dismantling the hexagonal tube by milling two opposite edges and spreading them apart, the pins are extracted in successive layers, sheared, and then dissolved in nitric acid at the start of the PUREX process. Twenty-six tons of Phénix assemblies were treated in this way, and 4.4 tons of Pu extracted (Appendix 2). It should, however, be noted that small adaptations will be necessary in the future for the treatment of MOX fuels richer in Pu, in particular for the management of insolubles.

For nitrides and carbides, the PUREX process also seems usable, after treatment of the residual sodium. Japan and Russia have studied this process for nitrides and, recently, India for carbides, with a few tons of spent fuel.

The application of the PUREX process to metallic FNR fuels would require, among other things, to take the following additional measures:

- Mechanical shearing of the pins under an inert atmosphere to avoid inflammation.
- Washing of the fuel with a neutron-transparent reagent to remove the sodium.

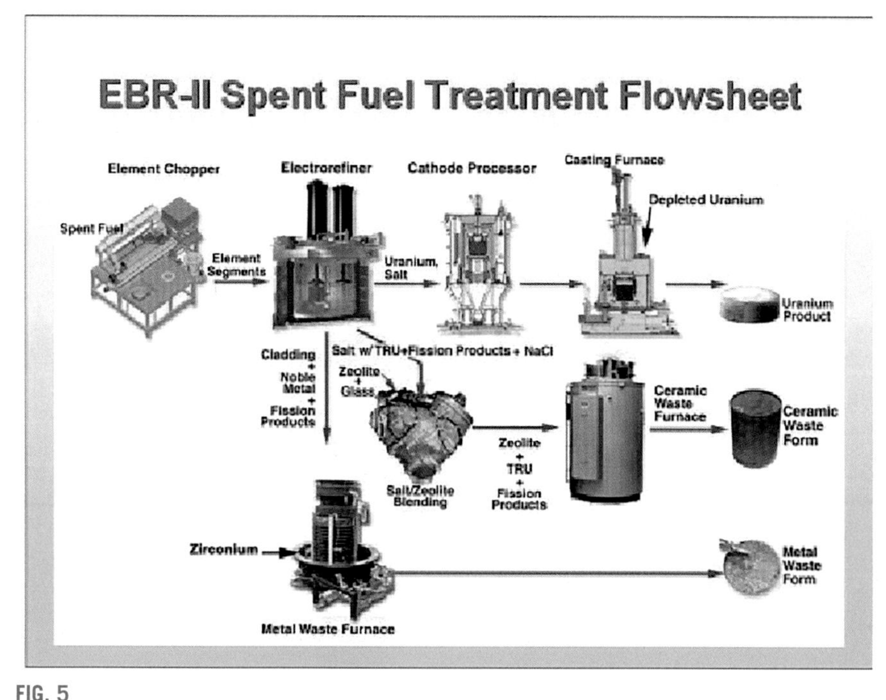

FIG. 5

View of the metallic fuel cycle developed on the site of the EBR-2 experimental reactor.

- After stripping, dissolving the fuel in a mixture of nitric acid and hydrofluoric acid under neutral sweeping to limit the hydrogen content in the sweeping sky below 4%.
- Compacting of solid waste under inert gas to prevent the ignition of any hydride deposits on the surfaces.
- Separate management of effluents containing fluorides.

It should be noted that the PUREX process has also been studied in India for the treatment of thorium oxides. Unlike uranium or plutonium, it requires the addition of a strong complexing agent, generally hydrofluoric acid, to catalyze the dissolution reaction. However, managing the fluoride ions in the process is difficult because of the corrosion problem of the process apparatus materials. For example, fluorides are banned at La Hague to prevent corrosion of stainless steel devices as well as the zirconium dissolver tank. The other problem is the management of uranium-233 generated by neutron activation of thorium-232. The problem is the decay of uranium-233 into protactinium-233, which is a strong beta-gamma emitter requiring remanufacture of the U fuel less than 6 months after the separation of uranium-233. India chose this research path because they do not have access to much uranium-235 (they are not a signatory of the Nuclear Non-Proliferation Treaty, NPT), but they have considerable thorium reserves (Fig. 6).

FIG. 6

View of the Indian site where the test, reprocessing, and fuel fabrication facilities are located.

A choice of fuel according to applications and strategy?

Nevertheless, the choice of type of fuel must also align with the type of solid fuel fast reactor that one wishes to develop, as well as the targeted strategy:

- In the case of fast reactors with lead coolant, the circulation speed of the coolant is low, because of corrosion problems, which requires a more aerated core than a fast reactor with sodium coolant. To compensate for the high surface fraction occupied by the coolant, the densest possible fuel is required, which tends to disqualify the MOX, the least dense of the four fuels. Metallic fuel also does not appear to be suitable owing to the high temperature target of lead coolant. In the end, the choice then falls on the nitrides.
- For a fast reactor with gas coolant (helium), studies, particularly those carried out at the CEA in the 2000s, showed the need for a dense, refractory fuel with a good conduction coefficient. Carbide was chosen following a critical analysis of feedback between carbide and nitride.
- India's long-term strategy is to use thorium, owing to its availability. Indeed, India has almost a third of the known thorium resources, i.e., around 500,000 tons, while it has only 1.5% of uranium reserves. Thorium being a fertile element, it is necessary, prior to its introduction into dedicated reactors, to use plutonium as a fissile element. This plutonium is obtained by processing the sodium-cooled FNR fuels that India is producing. To optimize plutonium production, India is striving to achieve high breeding rates, which would be facilitated by the use of a metal-alloy-type fuel. In general, the doubling factor is therefore a parameter to be taken into account. It could tilt the choice of certain countries toward metal fuel depending on whether they have sufficient reserves of Pu to start their industry.
- For some fast-spectrum small modular reactors, the current preferred option is to aim for long run times without recharging. From this perspective, metal fuel, which is denser, with a harder spectrum, is the first choice, especially in the United States.

Conclusion

Four types of solid fuels, that is, oxide, carbide, mixed U and Pu nitride ceramics, and the metal alloy UPuZr, remain potentially acceptable candidates as fuel for a sodium-cooled FNR even they differ in terms of feedback experience, level of qualification, and maturity.

In all cases, the maturity of the fuel cycle will have an influence on the choice of fuel for the type of fast neutron reactors to be deployed. This deployment will also depend on the availability and eventual needs of Pu to supply this type of reactor.

Each of these fuels has qualities and shortcomings both in operation and in manufacturing and reprocessing required to complete the cycle. However, the choice of a country is affected by historical reasons and general strategy.

According to the current state of knowledge, a pragmatic choice would be oxide, which has by far the greatest operating experience feedback in terms of reactor operation, manufacturing, and reprocessing (Fig. 7).

FIG. 7

View of the three oxide fuel elements tested on the Rapsodie, Phénix, and Superphénix fast reactors.

CHAPTER

Update on the technical state of sodium-cooled fast reactors in Europe

6

Abstract

France and Europe have developed extensive experience feedback on sodium-cooled fast reactors (SFRs), which culminated during the operation of Superphénix (1200 MWe) with the creation of a 1500 MWe European project called European Fast Reactor (EFR). The design of this project took into account feedback from the construction of Superphénix, to improve the design of the reactor. On this basis, from 2017 to 2022, the European ESFR-SMART project has been modifying the design to improve safety, taking into account the latest developments on the subject as well as studies from the French ASTRID project. The goal is to comply with the latest security demands following Fukushima.

In this context, the redesign is being carried out with simplicity and passivity in mind. An attempt has also been made to eliminate by design the occurrence of a number of known accidents through experience feedback.

A number of simplifications have been proposed and precalculated: elimination of the safety vessel, elimination of the dome or polar table, elimination of the decay heat removal systems inside the primary vessel, change to straight and shortened pipes for the secondary loops, etc. As a result, the final design is more compact and simpler.

Passivity has been introduced with secondary loops operating in natural convection, an additional control rod system capable of stopping the reactor without order from control command, decay heat removal systems operating in natural convection, and passive thermal pumps.

Finally, in terms of mitigation, a set of choices make it possible to ensure excellent mitigation in the case of a severe accident: integrated core catcher, thick slab ensuring containment, passive decay heat removal, etc.

In conclusion, these options have also been evaluated for lower-power SFRs, where they are also applicable and with threshold effects that allow certain additional simplifications to be implemented.

This chapter provides an overview of the technical state of SFR design in Europe in 2021.

Fast Reactors. https://doi.org/10.1016/B978-0-12-821946-1.00008-5

Keywords

EFR, ESFR-SMART, Passivity, Decay heat removal, Straight pipes, Natural convection, Practical elimination

The ESFR-SMART project

In 1988, while the European Superphénix reactor was in operation, a new European (sodium) Fast Reactor (E(S)FR) project, with a slightly higher power of 1500 MW_{el}, was launched in collaboration between France, Italy, Germany, and the UK. This project was discontinued with the shutdown of the Superphénix reactor. On the basis of a summary of the modifications explored in this project [1], a project called Collaborative Project on ESFR (CP ESFR), was initiated a few years later to "groom" EFR options and integrate the new technical developments [2]. It is on this new basis that a project called ESFR-SMART started at the end of 2017, with the main objective of integrating the new safety rules resulting primarily from the Fukushima accident [3].

The ESFR-SMART project is what in the Anglo-Saxon world is called a "working horse" or a "concept car." Its role is to introduce, outside constructive planning, new ideas for the future, which can be valuable guides for R&D. Unlike an "industrial" project, which works according to a construction schedule, here one can introduce innovative ideas, even if their lower level of technological readiness would require development and time. For these new ideas, research and preliminary calculations are performed to check their general feasibility and rule out major hurdles. The project is not designed to necessarily create solutions that can be readily used by a committed industrialist, requiring validation after numerous additional files submitted to the Nuclear Safety Authority (ASN), but rather, as mentioned earlier, narrows down further R&D directions to the most feasible and promising concepts.

In this sense, one of the main goals of the project was to select, implement, and assess new safety measures for a commercial-size SFR in Europe.

History of sodium-cooled fast reactors in Europe

In view of the potential of SFR, France, Germany, and the UK embarked on the construction of SFRs in the 1970s: the "Phénix" SFR in France, the Prototype Fast Reactor (PFR) in the UK, and the "KNK" SFR in Germany. The gradual phasing out of nuclear power in Germany and the UK in the 1980s led to the termination of fast-reactor development programs in these countries.

Only France, backed by the success of its reprocessing activities, continued along this path with the construction of the Superphénix reactor (1200 MW_{el}). This was already a European reactor, whose development involved strong collaboration with Italy (30%) and a Germany/Netherlands consortium (around 5%). As it was subject to strong opposition from environmentalist groups and a lack of political support, the

FIG. 1

View of Superphénix reactor (1987).

reactor was shut down in 1996 for purely political reasons, after a year of successful steady operation. Superphénix spent more than a third of its 10 years of operation able to operate but awaiting administrative authorization [4] (Fig. 1).

A new project on a smaller SFR of 600 MW_{el}, called ASTRID (Advanced Sodium Technological Reactor for Industrial Demonstration), was recently launched in France. This project enabled new advances in the field, in particular in the design of the reactor core. Sadly, the project was stopped in 2019, and, currently, no large- or medium-size commercial SFR project is planned in France or in Europe before the end of this century.

Given the huge potential advantages of these reactors, and their active development in countries such as Russia or China, Europe needs to maintain, with projects like ESFR-SMART, competence and an overview on what a SFR could look like at the end of this century. This view is supported by the fact that the pressure to minimize global warming could foster the reintroduction of efforts to implement new nuclear technology into the energy generation mix; thus, it is possible that the political interest will change in the future.

In this sense, the ESFR-SMART project is the best view available today of what a European fast reactor could look like in the future.

Safety improvement: Objectives and methodology

Since the previous CP ESFR project, the safety groups of the Generation-IV International Forum (GIF) have published new documents. In particular, a "task force" dedicated to SFRs proposed a set of rules to be applied to these reactors [5]. At the international level, the Fukushima accident in 2011 led to the issuance of new rules for all reactors [6]. These rules, not applied in the CP ESFR project, have been applied to the ESFR-SMART project.

The analyses and modifications proposed in ESFR-SMART are based on simplifications of the systems rather than adding new systems to the design, which is an important guarantee of safety. In general, safety authorities around the world tend to favor intrinsic and passive safety [7]. Many passive arrangements have, therefore, been introduced into the ESFR-SMART design to exploit the remarkable potential of SFRs in this field, including low-pressure fluid with good heat transfer and natural convection capacities.

In addition, the so-called practical elimination method was used to make impossible, by design, the occurrence of unacceptable incidents identified today for SFRs. More generally, significant feedback exists on the operation of the SFRs and on accidents that have occurred [4,8–10]. The ESFR-SMART design takes into account this feedback and valuable experience to both prevent such accidents or even make them impossible and to minimize their consequences.

Design was guided by the following objectives:

- Simplification of structures to enhance safety and improve manufacturing conditions for cost reduction and quality increase.
- Introduction of passive systems.
- Long duration (at least several days) before external intervention is needed.
- Improvement of in-service inspection and repair (ISIR) possibilities.
- Reduction of risks related to sodium fires and to the sodium-water reaction.
- Possibility to use the handling building for twin reactors (to reduce costs; the building layout is designed so that the handling building can serve two reactors).

The evaluation of the compliance of the reactor design with new safety rules provided the following recommendations:

- The loss of the DHR (decay heat removal) function should, by design, be practically eliminated on the basis of deterministic and probabilistic demonstration. This demonstration is managed by the use of diverse and redundant active systems in the pit (DHRS-3), by the possibility to use each secondary loop in an active or passive way (DHRS-2), and by six independent passive systems (DHRS-1) operating with the intermediate heat exchanger (IHX) even if the secondary loop is drained. The DHR function must be maintained in case of sodium leakages of the main reactor vessel.
- Passivity: additional passive systems are introduced, such as passive control rods operated without needing instrumentation and control (I&C) and electrical supply; natural convection capabilities of the primary and secondary system and of DHRS-1 and DHRS-2; thermal pumps able to passively ensure a flow rate or increased possibilities of operation in natural convection regime.

- Sodium fires: design measures are taken to avoid primary sodium leak above the roof even in case of severe accident with mechanical energy release. These measures aim at allowing a simplification of the design and avoiding dome or polar table in the primary containment. Measures are also taken to quickly detect and mitigate secondary sodium leaks.
- Prompt criticality risks: situations of large and rapid reactivity insertion likely to lead to prompt criticality must be practically eliminated. This notably concerns the risk of core compaction, which must be practically eliminated by a robust design and the possibility to monitor the core geometry in operation. Gas entrainment must be prevented by a careful design of structures, notably to minimize volumes of gas retention zones in the primary circuit.
- Mitigation of whole core meltdown: primary circuit must be mechanically robust, with a massive roof and a pit able to withstand sodium leakage. DHR systems as well as core catcher must not be sensitive to mechanical energy release (if any).
- Sodium-water reaction: secondary loop and steam generator designs must be robust to sodium-water reaction. The largest possible reaction must be intrinsically limited and demonstrated acceptable. Small modular steam generators should facilitate this demonstration.
- External hazards: new rules on external hazards are applied.

Some examples of safety improvement approach in the ESFR-smart

Reactivity control

New core concept with reduced sodium void effect

It is proposed to adopt a core with a globally zero or slightly positive sodium void effect, which contributes to reduce the consequences, in terms of potential mechanical energy release, of severe accidents [11].

Passive control rods

Passive control rods are proposed as self-actuated reactivity control devices for the core. The absorber insertion into the reactor is thus passively obtained, i.e., without the use of I&C, when certain criteria on physical parameters are met, e.g., low primary sodium flow rate or high primary sodium temperature [11].

Ultrasonic measurements to evaluate core geometry

As in the reactors already in operation, the pads at the subassemblies are forced into contact as power increases, leading to an increase in temperature. This prevents significant compaction of the core and reactivity increase during operation. Nevertheless, these measures are efficient if the pads are really in contact during the plant operation. Feedback experience on fast reactors shows that some subassemblies under irradiation swelling or bending effects can induce disturbance in the core geometry. Therefore, to practically eliminate significant possibility of core compaction, several measures are prescribed in ESFR-SMART to ensure that the pads are in contact. Ultrasonic measurements at the core periphery would make it possible to monitor global geometry during operations and verify the absence of significant changes to this geometry during operation.

Practically eliminated situations

More generally, preventive measures are to be taken for other practically eliminated situations related to the control of core reactivity, which include:

- Significant amount of gas passing through the core.
- Sudden deterioration of core support.
- Loading/unloading errors likely to have critical consequences for the core during handling operations.

Containment

Reactor pit taking over the functions of the safety vessel

The function of the CP ESFR safety vessel was to contain the sodium in the event of main vessel leakage, while maintaining in it a level of sodium sufficient to allow the sodium inlet into the IHX and keeping a sodium circulation cooling the core. To recover this function by the reactor pit (hence, suppressing the safety vessel), it is necessary to overlay the reactor pit with a metal-sheet liner to withstand possible sodium leaks and to bring it closer to the main vessel so that the volume between vessel and pit remains identical to the volume between the two vessels.

The replacement of the safety vessel by a liner with a DHR system attached gives the following anticipated advantages:

- Increase of the DHR capabilities through the reactor pit.
- Simplification of the safety demonstration with respect to the double leak of the main and safety vessels.
- A fault tolerant structure well adapted to the mitigation functions.
- A main vessel in-service inspection that remains possible, as the main vessel remains accessible from the reactor pit, by the top of the space between vessel and liner.

A special arrangement of the reactor pit is necessary to be able to operate in normal conditions, to support an accidental sodium leak of the primary vessel, and to cope with severe accident mitigation.

A reactor pit with a "mixed" steel-concrete structure (Fig. 2) is proposed for ESFR-SMART. A metallic liner is attached in front of the main vessel to support an oil cooling system. A material chemically compatible with sodium is provided between this "mixed structure" and the liner. This material must protect the mixed structure even in case of sodium leak through the inner sheet liner. This liner has no mechanical connection with the roof and, therefore, has more freedom for thermal

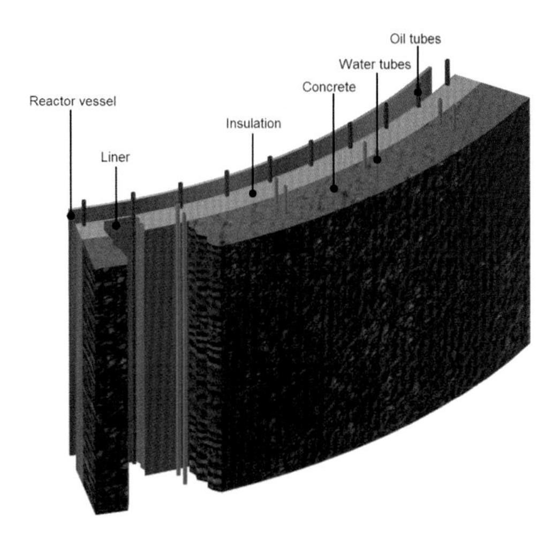

FIG. 2

Detail of the reactor pit with reactor vessel (RV), gas gap (GG), metallic liner (ML), sacrificial material (SM), concrete (C), and decay heat removal system (DHRS-3) in oil attached to the liner and in water inside the concrete.

expansion. Two reactor pit cooling systems are used. The first system is attached to the liner and very efficient in normal operation for residual power removal. The second system is installed in the mixed steel-concrete structure and should be able alone to maintain the concrete temperature under 70°C, even in case of severe accident.

A detailed description with thermal calculations of this pit organization is available in Ref. [12].

In-vessel core catcher

The mitigation of a severe accident with whole core meltdown will be achieved by means of a corium receiver, also called core catcher, located at the bottom of the vessel, under the core support structures, also called strongback (Fig. 3).

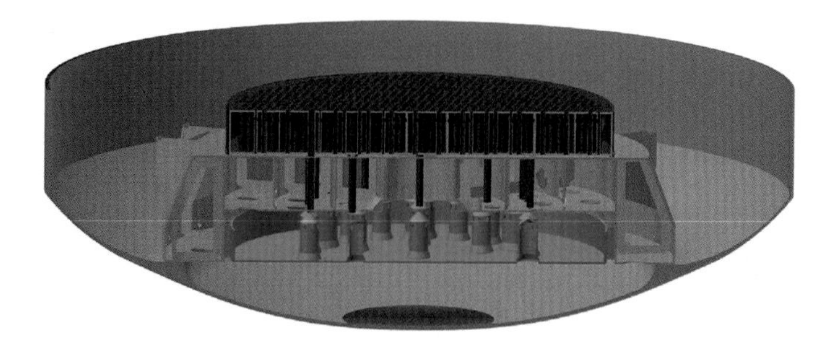

FIG. 3

View of the core catcher position inside the primary vessel. Vessel bottom (1), strongback (2), corium discharge tube (3), diagrid (4), and core catcher (5).

Transfer tubes, coming from the core, emerge above the core catcher to channel the molten corium. The use of molybdenum (as in BN-800; Ref. [13]), characterized by a high fusion temperature, is proposed for avoiding melting of the core catcher structure. Hafnium-type poisons can be used to avoid potential recriticality. The core catcher is designed for the case of whole core meltdown.

Precalculation of the core catcher has been provided [14] with calculation of the residual power of the melted core, and cooling of this core catcher by natural convection of the sodium around it (Figs. 4 and 5).

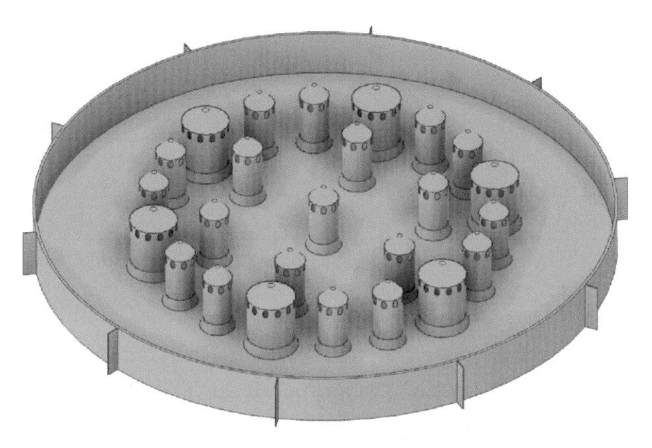

FIG. 4

Core catcher concept.

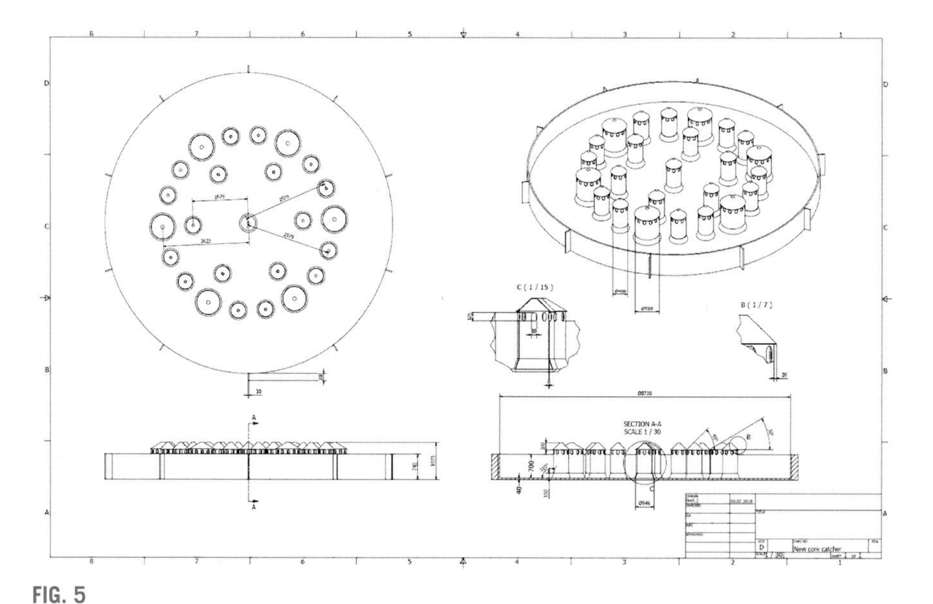

FIG. 5

Core catcher drawings.

Massive metallic roof

Superphénix experience feedback [8] suggests that the roof should be hot at its bottom part, to minimize sodium aerosol deposits. Moreover, to contribute to the practical elimination of large water ingress into the primary circuit, it is recommended by safety authorities to avoid water as roof coolant. The EFR massive metallic roof is therefore implemented, which presents many other advantages, including neutron shielding and mechanical resistance. Its thickness is defined by the industrial manufacturing contingencies, but should be about 80 cm. In the upper part, a heat insulator is installed to limit the heat flux to be evacuated during nominal conditions by air flow in forced convection or even natural convection.

Leak tightness of roof penetrations

It is proposed to study penetrations that have improved leak tightness during operation, with the goal of avoiding primary sodium leakage through the roof in case of mechanical energy release in a scenario of whole core meltdown. Such leakages are difficult to determine and can thus lead to conservative overpressure in the containment, making it necessary to implement systems such as dome or polar table, which are expensive and quite complex; prevention of such leakages could facilitate reactor operation.

To overcome these difficulties, the following modifications are proposed:

- For large components, pump and heat exchanger penetrations: they are already widely used to address earthquake issues. It is proposed to weld a sealing shell to ensure leak tightness in fast overpressure transient. These components are not intended to be handled frequently, but if handling is required, grinding will enable their easy removal.
- For rotating plugs: independently of the possible inflatable seals, eutectic seals, which are liquefied during the handling phases to enable rotation, are recommended to ensure leak tightness. Conversely, when operating the reactor, these seals are solidified, and the design retained should be such that there is no possibility of leakage in the case of a severe accident with mechanical energy release. These systems were used successfully on Phénix and Superphénix reactors with good feedback experience [4,9].
- To improve the primary sodium confinement in the main vessel, it is also proposed to consider:
    - an integrated primary cold trap, like in Superphénix, to limit the amount of primary sodium outside the vessel;
    - a sufficiently low argon pressure in the cover gas to avoid an accidental sodium-fountain effect of a defectuous plunging tube.

Decay heat removal

The secondary circuits are the nominal power removal circuits. They are very useful in DHR since they allow the creation of a cold column, in the IHX, essential for good natural convection in the primary circuit. The secondary circuit design is optimized to enable good heat removal by air in natural convection, such as in situations like "Fukushima" where both the cooling water and alternating current (AC) power supply are lost. For this purpose, several measures are taken:

- A secondary loop design enabling easy establishment of natural convection is adopted. Sodium leaks, inherent to a mechanical pump in operation, are recovered by gravity in the pump body toward the storage tank as in Superphénix. Sodium is purified at this level, and the purified sodium returns to the main circuit.
  The CP ESFR design for steam generators (SGs), with six modules per loop, is maintained. We take advantage of the large exchange surface, related to the SG modular design, to cool these modules by air in natural or forced convection (through hatch openings, like in Phénix reactor [9]). This is the heat sink for the secondary loop. We call this system Decay Heat Removal System 2 (DHRS-2).
- Finally, one or more thermal pumps are added in the secondary circuits. Thermal pumps are passive electromagnetic pumps using thermoelectricity provided by the difference in temperatures, with no need for external electricity supply. They provide a flow rate also in nominal conditions, and help to avoid stratification in the secondary loops, in case of loss of forced convection (Fig. 6).

In addition to these secondary DHRS loops, there are two independent cooling circuits in the pit: one in the reactor pit, with oil, installed on the liner, and one with water inside the concrete, capable of maintaining the entire pit at temperatures below 70°C. Suppressing the safety vessel makes these devices disposed in the liner much more efficient, with the goal to ensure a large part of the DHR. We call this system DHRS-3 (see thermal calculations of DHRS-3 in Ref. [15]).

To further reinforce the practical elimination of loss of DHR function, we add cooling circuits by sodium-to-air heat exchangers connected to the IHX piping.

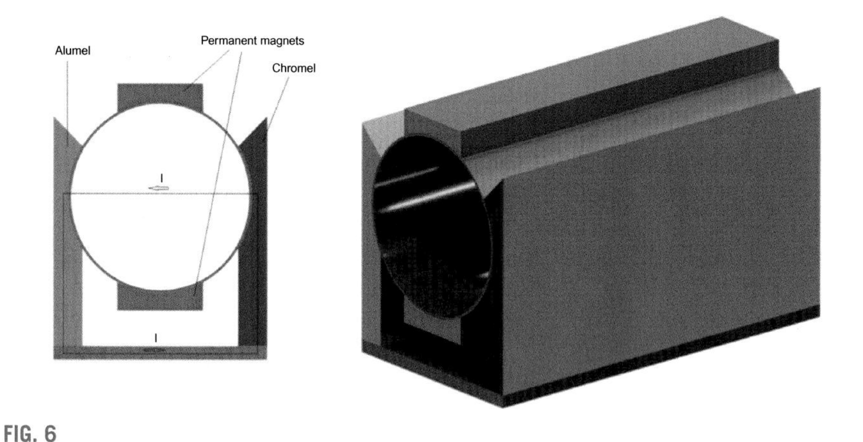

FIG. 6

ESFR-SMART thermal pump concept for secondary circuit: Alumel (1), Chromel (2), permanent magnet (3).

These circuits, which we call DHRS-1 (primary DHR system), have the following advantages compared with independent systems located in the primary circuit (formerly used in the CP ESFR design):

- No additional roof penetrations are required (increase of main vessel diameter).
- The cold column is maintained in the IHX, which guarantees good natural convection in the primary circuit.
- These systems are less sensitive to mechanical damage in case of severe accident, because they operate outside of the vessel.
- These systems operate in natural convection both for air and sodium.

This circuit DHRS-1 is available even when the secondary loop is drained, because the IHX remains always full of sodium. It operates in natural convection, but the addition of a thermal pump can further increase its capabilities and facilitate the starting of the operation.

Indeed, if the secondary circuit were drained, the secondary sodium in the IHX would be near atmospheric pressure. It is then necessary to place the sodium-to-air IHX below ~ 12 m (the sodium height corresponding to 0.1 MPa of atmospheric pressure) up the upper point of the IHX, to avoid cavitation (Fig. 7).

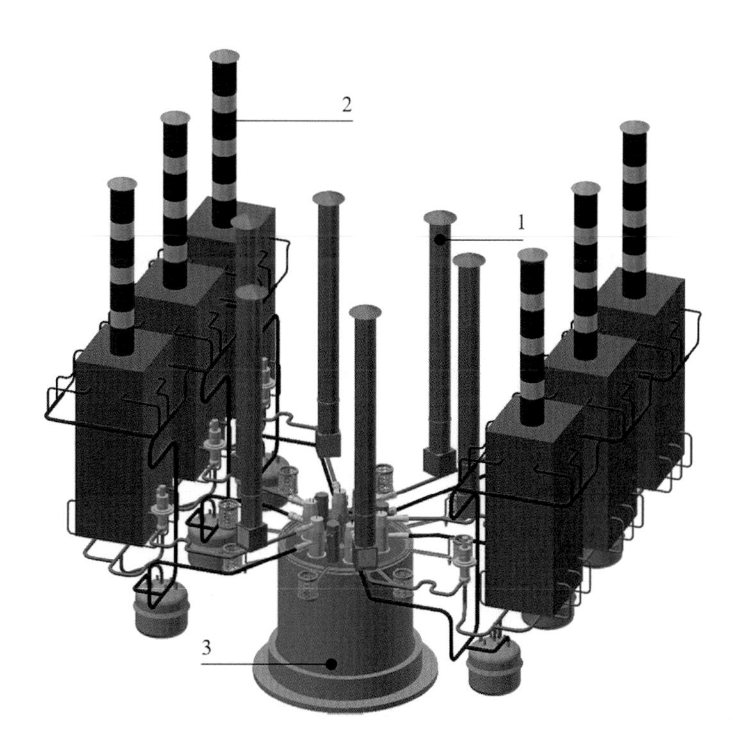

FIG. 7

ESFR-SMART main view of the reactor with the three DHR systems.

Thermal calculations of the three DHRS are provided: for DHRS-3, see Ref. [15]; for DHRS-1 and DHRS-2, see Refs. [14, 16]. For calculating DHRS-2 operation in natural convection, the code Cathare was used to determine the efficiency of the cooling by natural convection of air in the casings.

Sodium fire

The provisions to prevent leakage of primary sodium are explained above. Regarding the risks related to secondary sodium leakage, it should be noted that leakages are mainly a chemical risk considering that no or very little radioactivity is present in the secondary sodium circuit. Special dispositions are provided for easy detection and mitigation and are explained in the chapter on secondary loops.

Sodium-water reaction

Conventional devices enable one to control this risk efficiently and quickly. Modular SGs are retained, owing to their ability to quickly detect sodium water reaction and subsequently depressurize/isolate and drain the faulty module. The choice of modular SG allows also one to minimize envelope accidents. In case of sodium-water reaction, the objectives are to limit consequences on the plant operations such that operation can continue with remaining modules, after faulty module isolation. Risk of sodium-water-air reaction is mitigated with appropriate casing sizing.

A more robust design than CP ESFR is being studied to achieve severe accident mitigation, with the following objectives:

- A core catcher is provided at the bottom of the vessel, designed for whole core meltdown. Mitigation devices inside the core (corium discharge tubes) are intended to channel the molten fuel to the core catcher.
- The recriticality of the molten core should be prevented by using dedicated material such as hafnium inside this core catcher.
- The reactor pit should accept sodium leakage and, with its upper thick metal roof, form a solid, tight, and coolable containment system.
- The cooling of the primary circuit structure is achieved by DHRS even in case of severe accident.

Dosimetry and releases

It is known that, during SFR normal operations, the radiological releases are almost zero for gas. Release of primary argon, which may occur in normal operation, is sufficiently delayed to reduce radioactivity. The only liquid radioactive release is the liquid used to wash fuel subassemblies or wash and decontaminate components [4,8]. In terms of personnel dosimetry, this reactor design leads to a dosimetry much lower than in pressurized water reactors (PWRs) [17]

Indeed, the work sites generating the highest doses in PWRs either do not exist in SFRs or exist but generate very low doses, because the primary circuit is entirely contained in a vessel and the secondary circuits are not active; in PWRs, doses are high, for example, in the reactor vessel opening operation (which do not exist on SFRs) and in the removal and installation of thermal insulation (a frequent operation on SFRs but with negligible radiological doses).

Other elements also explain this difference in dosimetry. For example:

- In SFRs, primary sodium purification cartridge handling and processing are conducted under biological protection.
- In SFRs, primary components are decontaminated prior to the maintenance operations they must undergo.
- SFR secondary circuits are not radioactive.

This benefit will be kept for ESFR-SMART.

Simplicity and human factor

Starting from the CP ESFR design [2], our approach has consisted in proposing the simplest possible reactor, while keeping the necessary lines of defense. It is expected that this simplicity should contribute to reactor safety, by making it easier to operate. Compared with CP ESFR, the following simplifications are proposed:

- Dome (or polar table) suppression.
- Safety-vessel functions taken over by the reactor pit.
- Primary sodium containment improvement.
- Natural convection cooling enhancement on the secondary side.
- Optimized and simplified DHRS-dedicated circuits.

Passive and redundant systems that are independent of I&C or of the operators' action will enable the reactor reactivity control and its cooling by natural convection, even in the most severe cases of simultaneous loss of cooling water and electrical power supply. With these improvements, the new design is more forgiving, with respect to the reactivity control as well as at the intervention time required from the operator (enhanced grace period).

Description of ESFR-SMART primary system including these new options

General plant characteristics

The 1500 MW_{el} reactor is of pool type and based on several key design options aiming at a high level of safety, robustness, and manufacturability of all components, including:

- A massive metallic reactor roof (~ 80 cm thickness), limiting heat fluxes, providing good dosimetry behavior, and able to withstand any accidental situation. This design was already proposed for the EFR [1]. In normal operation, the roof is hot in the lower part (to avoid sodium deposition) and actively or passively cooled by air in the upper part.

- A diagrid with two primary sodium inlets (LIPOSO) for each pump.
- An inner vessel with a conical part (redan) with 25 degrees slope.
- An internal core catcher of high capacity located under the diagrid and strongback.
- A safety liner on the reactor pit surface able to withstand sodium leak and to mitigate any accidental situation.
- An inner storage of spent fuel near the core.

Other components are based on classical designs:

- An above core structure (ACS) to guide control rods, support instrumentation, and reduce the high sodium velocity at core outlet.
- Two eccentric rotating plugs to take any of the fuel assemblies by means of a direct lift charge machine (DLCM) and a fixed arm charge machine (FACM).
- Six IHXs.
- Three primary pumps (PP).
- A DHR system (DHRS-3) integrated into the reactor pit.

A general view of the primary system is shown in Fig. 8.

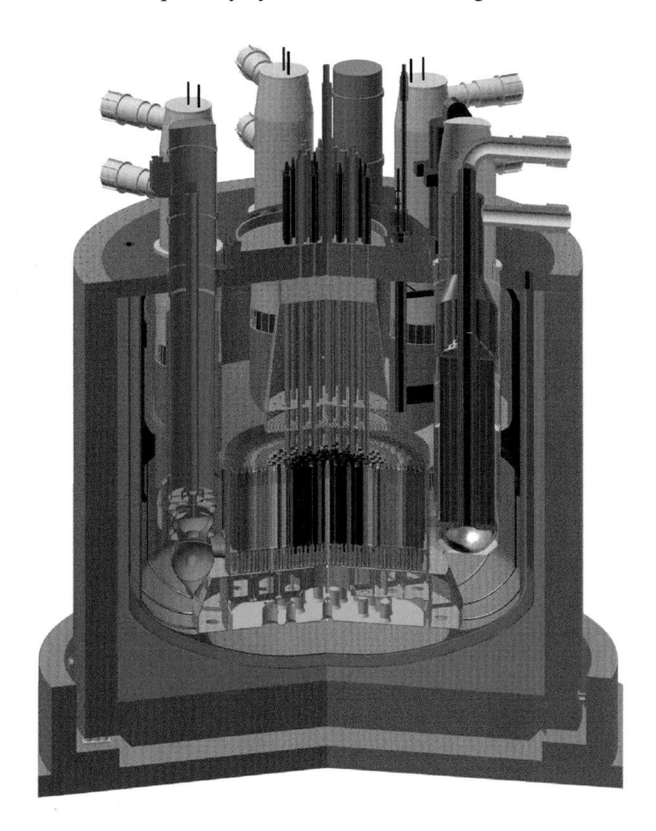

FIG. 8

View of ESFR-SMART primary system.

Main parameters of the ESFR-SMART plant are given in Table 1. The following considerations were taken into account when selecting the main temperature parameters of the plant:

1. Core inlet and outlet temperatures (395–545°C): this choice provides a substantial logarithmic mean temperature difference for the IHX. The core temperature increase of 150°C is a compromise between plant capital cost, thermal loading considerations, material behavior, and pumping power requirements.
2. Secondary sodium temperature at SG inlet (530°C): this is a compromise between acceptable material properties and minimization of surface area requirements. Moreover, this temperature seems a reasonable upper limit for the secondary hot leg temperature.
3. Secondary sodium temperature at SG outlet (345°C): this choice provides acceptable temperature differences and limits the temperature difference between sodium and water on the lower tubular plate of the SGs.
4. Steam temperature (528°C): this choice is connected to the creep properties of the steam tube and tube plate materials of the SG.

Table 1 Main parameters of the plant.

General	
Thermal power (MW_{th})	3600
Net electrical power (MW_{el})	1500
Global efficiency (%)	42
Plant lifetime (years)	60
Availability target (%)	90
Mass of sodium in main vessel (t)	2350
Total pressure losses in primary system (MPa)	0.45
Cover gas above primary sodium free level	Argon
Pressure of cover gas (MPa)	0.115
Core	
Core inlet/outlet temperatures (°C)	395/545
Type of fuel	$(U,Pu)O_2$
Fuel enrichment (%)	17,99
Core global geometry	Cylindrical, 3 layers of reflectors
Core Support	Strongback resting on primary vessel bottom
Core mass (t)	~430
Core outside diameter (m)	~8
Core flow rate (kg/s)	18,705

Continued

Table 1 Main parameters of the plant.—cont'd

Core bypass flow rate (kg/s)	831
Sodium supply	Pump connection to diagrid
Diagrid and core support materials	316 L(N)
Diagrid mass (t)	~70
Sodium leak in diagrid for cooling systems (kg/s)	830
Core pressure drop (including inlet and outlet) (MPa)	0.38
Core support pressure drop (diagrid) (MPa)	0.07
IHX	
Number of IHXs	6
Power of one IHX (MW)	600
Type	Tubular, counterflow
Material	Stainless steel
Pressure loss (primary) (MPa)	0.025
Working fluids, primary/secondary	Sodium/sodium
Mass of one IHX (t)	127
Primary sodium temperature at IHX inlet/outlet (°C)	545/395
Secondary sodium temperature at IHX inlet/outlet (°C)	530/345
Primary pumps	
Number of primary pumps	3
Type	Mechanical, radial admittance, axial exhaust, antireverse-flow diode
Mass of one pump with motor (t)	~164
Location	In reactor vessel
Nominal rotational speed (rot/min)	450
Net positive suction head available (m)	13
Pressure head (MPa)	0.45
Nominal flow rate (kg/s)	6512
Halving time (LOSSP) (s)	~10
Minimum time from 100% to 25% of nominal speed (s)	30
Secondary loops	
Number of secondary loops	6
Composition	1 IHX, 6 SGs, 1 secondary pump, 1 thermal pump, 1 purification system, 2 draining systems
Nominal flow rate per loop (kg/s)	2541
Length of pipes with Ø 0.850/Ø 0.350 m per loop (m)	~193/~30
Mass of secondary sodium per loop (t)	~254

Table 1 Main parameters of the plant.—cont'd

Steam generator	
Number of steam generators per secondary loop	6
Type	Modular, tubular, counterflow
Mass of one SG (t)	~50
Material	9Cr–1Mo modified
Working fluids, secondary/tertiary	Sodium/water
Power of one SG (MW)	100
Water inlet/steam outlet temperature (°C)	240/528
Steam pressure (MPa)	18.5
Steam flow rate per secondary loop (kg/s)	287
Vessels and structures	
Main vessel height/diameter (m)	17.185/17.56
Main vessel material	Stainless steel 316 L(N)
Main vessel mass (t)	~900
Main vessel cooling	With cold sodium taken from diagrid, immersed weir
Main vessel cooling mass (t)	~80
Cold/hot sodium separation	Inner vessel with conical part (redan)
Redan mass (t)	~200
Safety liner mass (t)	~284
Reactor roof type	Massive, hot in the lower part and cooled by air in upper part
Reactor roof material	Stainless steel 16MND5
Reactor roof mass (t)	~1300
Total mass supported by reactor roof (t)	~7300
Above core structure type	Conical
Above core structure mass (t)	~550
Decay heat removal system-1	
Number of DHRS-1 loops	6
Material	Stainless steel
Mass of one loop (t)	~20
Fuel handling	
Fuel handling mechanism	2 rotating plugs (1 eccentric) in reactor roof
Mass of each rotating plug (t)	~200
Spent fuel extraction from reactor	Gas flask
Spent fuel transfer	With in-reactor fuel handling station
Intermediate storage	External storage in water pool

Core

Compared with the core defined in the CP ESFR project [2], the following major modifications are implemented in the axial layout of the core to reduce reactivity effects in case of sodium boiling under hypothetical accident conditions: (a) a large sodium plenum topped by absorber is introduced above the core, (b) the inner/outer core heights are reduced by 25 cm/5 cm, and (c) a lower fertile blanket is introduced below the fuel, with a steel blanket below. The Pu content is the same in the inner and outer core zones. The full specification of the new core can be found in Ref. [11] (Fig. 9).

The measurement of the core dimensions with ultrasound sensors is expected to be achievable, and should take into account uncertainty measurements during the reactor lifetime. Devices of this kind had been installed at Phénix after the observation of some negative reactivity, to monitor the core geometry.

Concerning the maturity of the passive control rods, several types of passive control rods are being studied or even tested [11]. For example, the triggering of Curie-point rods at a certain temperature seems feasible and is currently being tested. Other types (differential thermal expansions, etc.) do also exist. Passive hydraulic control rods were already installed in the BN-800 reactor in Russia [13].

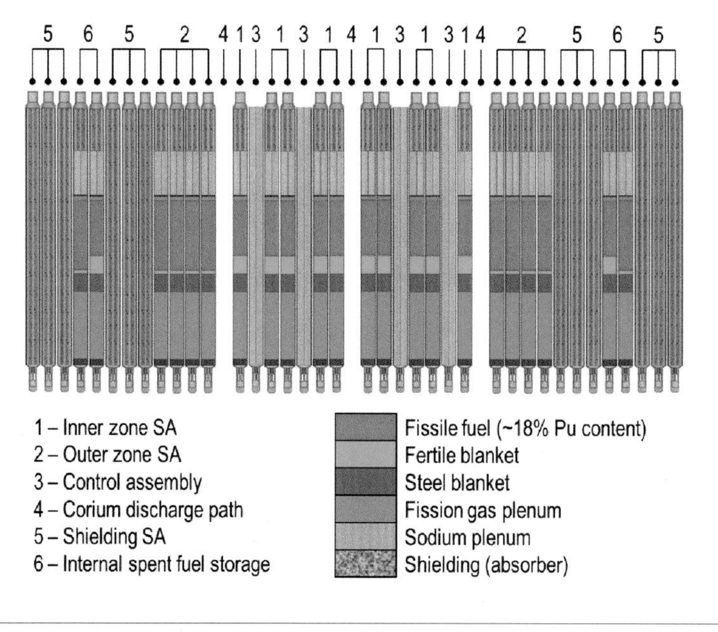

FIG. 9

Core axial map.

Main vessel

The main vessel has a diameter of 17.56 m and a height of 17.185 m. It is taller than EFR (16 m high) because the sodium level is 1 m higher, mainly to increase the IHX heat exchange length. The main vessel is fabricated of austenitic steel (316LN) and hung on the roof by means of a forged piece. The upper cylindrical part is cooled by a small sodium flow taken from the cold plenum below the diagrid. An immersed weir limits the risk of gas entrainment and ensures creep and fatigue resistance of the main vessel over 60 years.

Inner vessel

The inner vessel is an asymmetrical shell fabricated structure, comprising:

- A conical part (redan).
- A cylindrical inner vessel upper skirt.
- A cylindrical core barrel welded on the strongback.
- A minimum set of nine penetrations in the conical skirt: six for IHXs and three for PPs.

The inner vessel separates the hot pool, which contains the core subassemblies and the IHX inlets from the cold pool where the IHX outlets and the PP inlets are located. It provides a leak-tight barrier between the hot and cold pool and provides a geometric and hydraulic guide to the pump inlets. It serves together with other internal structures to distribute the primary sodium flow inside the main vessel.

A conical skirt (redan) is proposed in place of the oval skirt of EFR to simplify manufacturing. The core barrel is welded on the strongback (not on the diagrid). As a result, the conical skirt is less sensitive to buckling by plastic damage. It also decreases the surface subjected to the difference in pressure between collectors, and provides natural protection of the diagrid against hot shocks at the exit of the IHX.

As the upper part of the inner vessel is subject to high thermomechanical constraints at the free surface (sodium at 545°C, argon at 400°C), the wall thickness in that region is minimized to lower thermal fatigue. On the other hand, it must also sustain high mechanical and fluid loads during an earthquake, so its thickness must be large enough to ensure its stability.

Reactor roof

A massive steel reactor roof is proposed, following design and feasibility studies of EFR [1]. It has several advantages over the conventional fabricated box structure filled with concrete of Superphénix:

- No water inside the roof.
- Good dosimetry protection.
- Low heat flux by conduction.
- No sodium deposit in the hot lower part.
- Very good mechanical behavior even in the case of severe accidents.

The roof is 0.8 m thick, made of the same steel grade as a PWR vessel, and fabricated in sectors using narrow gap welding. All welded joints can be controlled. Component penetrations are designed to reduce tolerances, limiting heat transfer from the gas cover (compared with conventional designs with concrete). Access to the top surface of the roof could be made possible by extra sheets of insulation material, but in case of loss of power supply, natural cooling must be optimized to maintain an acceptable roof temperature. The roof lies on the mixed concrete/metallic structure of the reactor pit that also supports the main vessel weight. Access to circumferential welded joints is made possible by a series of regularly spaced holes.

The rotating plugs are also solid and made of stainless steel; thus, the design of the heat transfer constraints is consistent for the whole reactor roof. Since eutectic seal reduces its volume when frozen, the plugs are lowered for normal reactor operation, so that the thermocouples that monitor the core can be set closer to the subassembly heads to provide a more representative measurement during power operation.

A mechanical seal ensures the roof leak tightness at component penetrations (IHX, PP), and a supplementary ring around these components is welded to ensure total leak tightness during reactor operation, even in the most severe accidents. With the same objective, a frozen eutectic seal ensures the tightness of the rotating plugs during reactor operation (as was done in Phénix and Superphénix).

Concerning the degree of maturity, this type of a roof meets all requirements perfectly. The industrial feasibility of tall welds could, however, require reduction of the thickness. It is then possible to add, in multiple layers, heat insulation with metallic thickness.

Reactor pit

The reactor pit is a concrete/metallic structure supporting the roof. The safety vessel between the pit and main vessel is proposed to be suppressed. A metallic liner covers a structural insulation between the main vessel and the concrete/metallic structure (Fig. 4).

In the unlikely event of a leakage of the main vessel, hot sodium flows into the pit. The liner should prevent contact between hot sodium and insulation and provide improved thermal exchange for cooling. The insulation is chosen to be chemically compatible with sodium. The volume available for the sodium between the liner and the main vessel is calculated to maintain the sodium circulation between the heat exchangers and the core (Fig. 10).

DHRS-3

The concrete/metallic structure of the reactor pit is cooled during normal operation by the pit cooling system. Two independent active cooling systems are proposed in the reactor pit (we use the acronym DHRS-3 for the combination of these two systems):

- The oil cooling system (DHRS-3.1) is installed in the gap between the insulation liner and the reactor vessel or inside the insulation as shown in Fig. 4. The oil system is in direct view of the main vessel and therefore more efficient without the screen of a safety vessel. The oil under forced convection

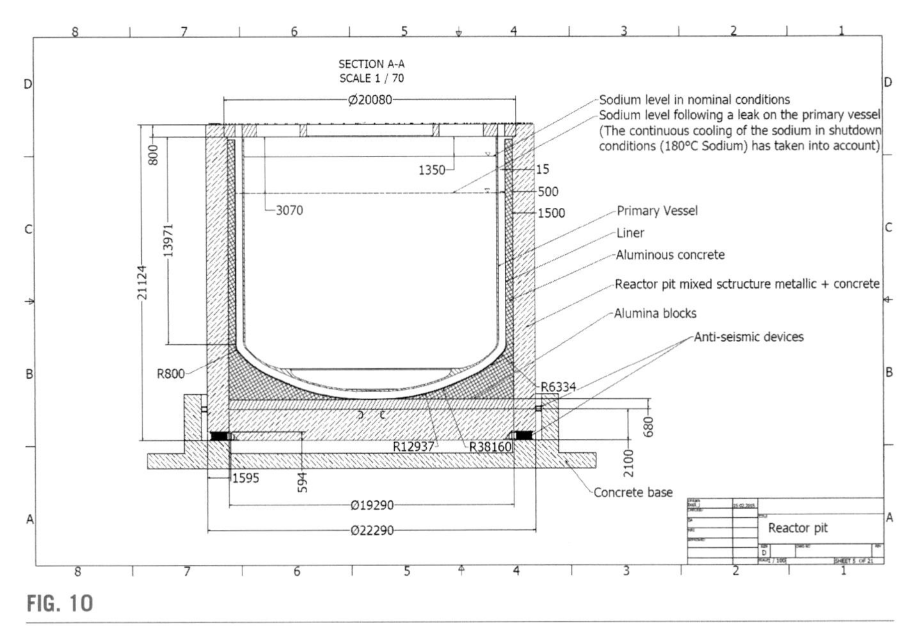

FIG. 10

ESFR-SMART pit drawings.

can remove the heat transferred by radiation from the reactor vessel at high temperature. Unlike water, oil is able to operate at high temperature.

- The water cooling system (DHRS-3.2) for the concrete cooling is installed in the concrete/metallic structure and aims to maintain the concrete temperature under 70°C in all situations, even if the oil system is lost.

Both oil and water circuits work during normal operation and must maintain the concrete temperature below 70°C. After reactor shutdown, the oil system alone must be able to remove all the decay heat generated in the reactor, after a certain time delay (3 days). In case of reactor vessel leak and loss of the oil system, the water system should be able to remove all the decay heat generated in the reactor.

Primary sodium confinement

A number of measures have been taken to ensure the primary sodium confinement:

- Tightness measures already described.
- A reactor pit able to withstand sodium leakage.
- No circulation of primary sodium out of the vessel. The purification circuits are components integrated into the vessel (as in Superphénix).
- Low pressure of cover gas (argon), to avoid "geyser effect."
- A massive roof to withstand severe accident.
- Slight overpressure in the secondary loops to avoid primary sodium leaks inside the secondary loop.

All these measures aim at avoiding overpressure due to a primary sodium fire; thus, dome or polar tables are avoided, which are very large components (about 15 m high), are quite expensive, and make handling operations difficult for the reactor operator. This is a major simplification of the design and, thus, a safety improvement.

Core support structure and connection to pump

The core support structure, including strongback and diagrid (Figs. 2 and 5), must be designed to avoid damage.

The lower part of the reactor vessel is shown in Fig. 5. The core and neutron shielding are supported by a diagrid, which rests on the strongback. The strongback is laid on the main vessel bottom and transfers the total core and diagrid weight.

The diagrid is a stainless steel cylindrical structure of 8 m in diameter containing a large number of vertical circular shroud tubes into which the core subassemblies are inserted. These shroud tubes provide the positioning and support for the subassemblies and allow the sodium feed from the diagrid through holes. The detailed flow arrangements are such that the hydraulic forces acting on the subassemblies serve to hold them down in the diagrid.

The diagrid structure is entirely welded. The absence of bolts removes any risk of loose parts inside the primary circuit. The diagrid structure needs high stability (thermal and mechanical) to avoid changes in core geometry. The pumps are connected to the diagrid, by two tubes, like in Superphénix.

The strongback is a stainless steel box-type structure comprising two circular plates linked by welded webs. It is resting on the vessel bottom, supports the diagrid, and allows the feeding of cold sodium to the primary vessel cooling system.

Thirty-one discharge tubes come from the core and cross the diagrid and strongback to arrive above the core catcher. These tubes are filled with sodium in normal operation. Their function is, in case of severe accident with core melting, to provide a natural way for the corium to escape to the core catcher.

Core catcher

The mitigation of a severe accident with whole core meltdown will be achieved by means of a corium receiver, also called core catcher, located at the bottom of the vessel, under the core support plate, called a strongback (Fig. 5). The goal is to design the core catcher for the whole core fissile inventory. Corium transfer tubes, coming from the core, emerge above the core catcher to channel the molten corium. Chimneys are designed under these tubes to allow good dispersion of the corium inside this core catcher (Figs. 6 and 7). The use, as in the Russian reactor BN-800 [13], of molybdenum, characterized by a high melting temperature, avoids melting of the core catcher structure.

The use of hafnium-type poisons inside the discharge tubes is possible to increase the margin to recriticality. General design of the strongback structures is made to improve the natural convection of the sodium around the core catcher (see calculations in Refs. [16, 18]).

Primary pump

The PPs are mechanical pumps for which there is good feedback on experience and reliability [4,8]. The selection of three primary sodium pumps is consistent with the compact primary circuit layout incorporating six IHX, which enables the vessel size to be minimized. These three pumps are utilized in the primary circuit, located on a common pitch circle diameter, and form a parallel circuit with the same pump head and the same flow. The design includes the following main features, which are similar to the Superphénix pumps:

- Single mixed flow impeller.
- Top inlet entry flow to the impeller.
- Subcritical hollow drive shaft, designed to get the first critical whirling speed above the maximum operating speed with a comfortable margin.
- Synchronous motor to allow easy operation of the pump over the whole range of required speeds (with specific regulations).

The PP comprises a cylindrical casing, vertical shaft machine inserted into the primary circuit via a penetration in the reactor roof, on which it is mounted. The upper part of the pump is firmly connected (to avoid displacement during seismic events) and even welded to avoid sodium leakage during any accidental event. To allow radial displacement between the upper part that remains cold and the lower part at hot sodium temperature, the lower part of the pump can bend to accommodate differential thermal expansion. The impeller is mounted on the drive shaft. The advantages and disadvantages of implementing a flywheel to extend the rundown time must be assessed. The shaft is supported at the top end by a magnetic axial thrust bearing and a radial bearing. A hydrostatic sodium radial bearing is located at the bottom of the shaft above the impeller.

Intermediate heat exchanger

A simple straight tube design is proposed for the IHX. This is a counterflow heat exchanger, with the secondary sodium flowing downwards through a central duct before turning upwards in the bottom header and flowing vertically inside the heat exchanger tubes, where the primary sodium flowing downwards on the shell side heats it. Rated unit power and thermal cycle are in accordance with the specified performance (600 MW_{th}). This design is a proven version used in all SFRs worldwide, taking into account the Phénix and Superphénix experience feedback [8,9] and the need for an efficient mixer in the secondary sodium outlet recovery box after heating.

The IHX unit is connected physically and functionally to both the primary and secondary sodium circuits and as such must be designed to withstand the specified maximum secondary circuit pressure coming from the SG unit design-based accident (about 5.5 MPa on the IHX in case of a sodium-water reaction resulting from the failure of all the tubes of a modular SG).

The IHX has a valve on the primary sodium side allowing the inlet window to be closed and the secondary circuit and steam plant to be isolated from the primary circuit. This allows the reactor to be operated with one or two IHX isolated. The IHX is firmly connected on the roof (to avoid seismic displacements) and even welded to avoid sodium leakage. Thus, a seal between the IHX and the inner vessel must accommodate thermal expansions between the components and the conical part of the inner vessel (redan). A gas seal is not recommended to avoid risk of gas entrainment into the core, and a dedicated oscillating mechanical seal is considered.

The IHX must have a tube in the upper part to extract the hot secondary sodium for the DHRS-1 and a tube for the cold secondary sodium returning to the central part of the IHX. To take into account these new measures introduced in ESFR-SMART, the IHX diameter was increased by 10% compared with the EFR IHX.

Decay heat removal concept

While in the CP ESFR design there were dedicated DHRS in the primary circuit, in the ESFR-SMART design there are no dedicated DHRS in the primary circuit, neither in the cold pool nor in the hot pool. Indeed, the best way to have good natural convection level through the core is to always remove heat from IHX and therefore maintain in the IHX the cold column of the primary sodium. Following this concept, the heat removal possibilities of the secondary loops are used to the maximum extent possible, including in cases where the secondary sodium operates in natural convection and where the feedwater supply is lost. If the secondary loops are completely lost and drained, special systems (DHRS-1) maintain the cold column of the primary sodium in the IHX, which allows good core cooling.

Another advantage of the proposed measures is a reduction in the number of sodium circuits (each circuit should have the draining and purification systems). The proposed design includes only the secondary sodium circuits that are used for both DHRS-1 and DHRS-2. This simplifies the work of operators. The only DHR system in the reactor itself is the DHRS-3, located inside the reactor pit. This system is more efficient than in Superphénix, owing to suppression of the safety vessel and the direct view of the primary vessel by the oil cooling system (DHRS-3.1), without the screen of the safety vessel. A second system with water (DHRS-3.2) is mainly used for mitigation situations.

In summary, the DHRS operation is as follows:

- If feed water is available, one secondary circuit is enough to remove decay heat from the core at acceptable temperatures.
- If feed water is lost, the opening of the windows in the SG modules creates a passive heat sink by natural convection of the atmospheric air (DHRS-2).
- If all the AC power supplies are lost, the taken measures ensure good natural convection, including of the secondary sodium assisted by a thermal pump [14,16].
- Studies have minimized as much as possible the common modes that can lead to a simultaneous loss of all six loops. If secondary loops are nevertheless

drained, DHRS-1 connected to the IHX enables the passive removal of decay heat by natural convection of the atmospheric air and natural convection of the secondary sodium assisted by a thermal pump.

- Furthermore, redundant circuits diversified and secured at the reactor pit (DHRS-3) enable further cooling of the primary set, including in mitigation cases. This system alone can ensure the DHR for several days.

Polar table or dome

Neither polar table nor dome is considered in the ESFR-SMART design. Specific measures have been taken to avoid leakage of sodium in the primary building and, thus, overpressure due to a sodium fire.

Description of ESFR-SMART secondary loops

General description of ESFR-SMART secondary loop

The secondary system transfers the heat from the IHXs to the SG units during nominal operation and during operational DHR via the tertiary water/steam system after reactor shutdown. It represents a nonradioactive barrier between the radioactive primary sodium system and the non-radioactive tertiary water/steam system.

Each of the six loops has one IHX, one secondary pump, and one SG unit with six modules. Each of the loops has also one DHR system (DHRS-1) able to keep heat removal from IHX even if the main secondary loop and one purification circuit are drained. A quick draining system allows the sodium of the loop to be drained to the storage tank dedicated to this loop.

The sodium flow rate in each loop is 10,800 m^3/h (2541 kg/s), and the power removed by each of the SGs is 100 MW_{th} (Fig. 11)

Note that there is a permanent flow of sodium in the loop for purification and monitoring. The sodium leaks at the secondary pump (hydrostatic bearing and impeller labyrinths, both a few percentages of the flow) and is collected passively by gravity and sent to the storage tank. There, a lift pump returns this flow through a purification circuit to the loop. The permanent circulation of sodium is passively ensured in the DHRS-1, which therefore does not require sodium monitoring or purification circuit.

Secondary pump

The operational experience of the secondary pumps shows high reliability of these components and only a few problems in operation [7,8]. Tests of high-power electromagnetic pumps were carried out in the frame of the ASTRID project in France in 2017/2018 to assess the advantages/disadvantages of these systems compared with usual mechanical pumps. To date, no clear advantages have been demonstrated in

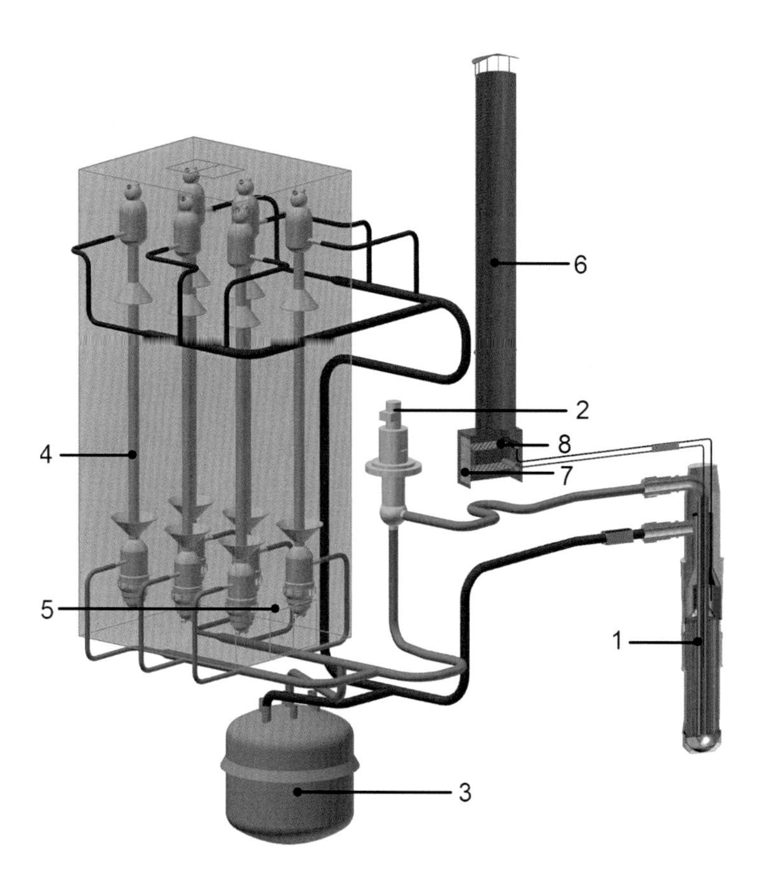

FIG. 11

General view of the ESFR-SMART secondary loop with the initial option of flexible pipes between the components.

terms of the cost, sizing, or reliability of these new systems. At the ESFR-SMART level, therefore, the EFR-size mechanical pumps were kept.

The dimensions of the secondary pump are given in Fig. 3. One potential issue with mechanical pumps is the sodium leakage in operation with the hydrostatic bearing and with the impeller seals. These leakages arrive to the cylindrical casing of this pump with a free surface of sodium. At this level, the leaked sodium is passively (by gravity) drained to the storage tank through a pipe. Similarly to Superphénix, this sodium, after purification, is sent back by the purification circuit to the secondary circuit. Even when the pump is stopped, a little leakage always exists owing to the difference of the levels of the pump free surface (at low point)

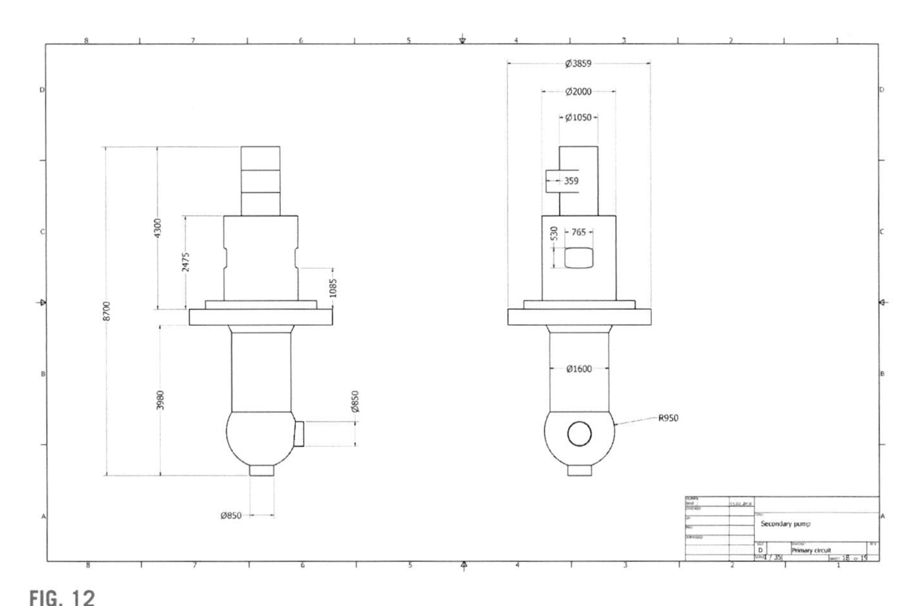

FIG. 12

View of the mechanical secondary pump.

and the SG free surface. Therefore, electromagnetic pump of the purification system must always be in operation to avoid unexpected draining of the loop when the secondary loop is stopped. The possibility of using batteries as the power supply for this pump will be assessed to ensure mechanical pump leakage recuperation even when all AC power supply is lost. Passive thermal pumps should also be proposed (Fig. 12).

Steam generator

The modular, straight-tube option was kept for SGs, which was also the EFR initial option. This type of module is used in the Indian Prototype Fast Breeder Reactor [9] and Russian BN-800 reactors; furthermore, it has been selected in several projects, including Prototype Generation-IV SFR (PGSFR) and Chinese fast reactor projects.

This modular option is preferred over the concept of the helical SG, selected and used for Superphénix [10], owing to several advantages regarding industrial manufacturing, maintenance (possibility of periodically replacing a module), and safety files (better detection and mitigation of accidents with sodium-water reactions). Moreover, having six modules in a loop makes it possible to increase the available heat exchange surface and, therefore, to remove the decay heat by simple natural convection of the atmospheric air around the walls of these modules

The main geometric characteristics of a 100 MW_{th} module are given in Table 2.

A global view of the SG module is shown in Fig. 4.

Table 2 Main geometric parameters of the steam generator.

Parameter	Value
Number of heat exchange tubes	364
External diameter/wall thickness of the exchange tube (mm)	15.6/2.5
Active/total length of the tube (m)	26.4/26.8
Outer shell diameter/wall thickness at the level of tubes (mm)	750/28
Outer shell diameter/wall thickness at the level of collectors (mm)	1900/40
Diameter of inlet, outlet, and discharge pipes (mm)	350
Tube material (chosen because of low thermal expansion coefficient)	[illegible]
Total length of SG (m)	29.1

The six modules are hosted in a casing with the possibility to open their windows and, with the help of a dedicated chimney, establish an appropriate level of natural convection of the atmospheric air. Sodium and water circuits are separated inside the casing; nevertheless, adequate prevention and mitigation of chemical releases in the environment in case of sodium leak should be further studied. Moreover, the casing should be able to support an airplane crash, to avoid fires from mixed air, water, and sodium (Fig. 13).

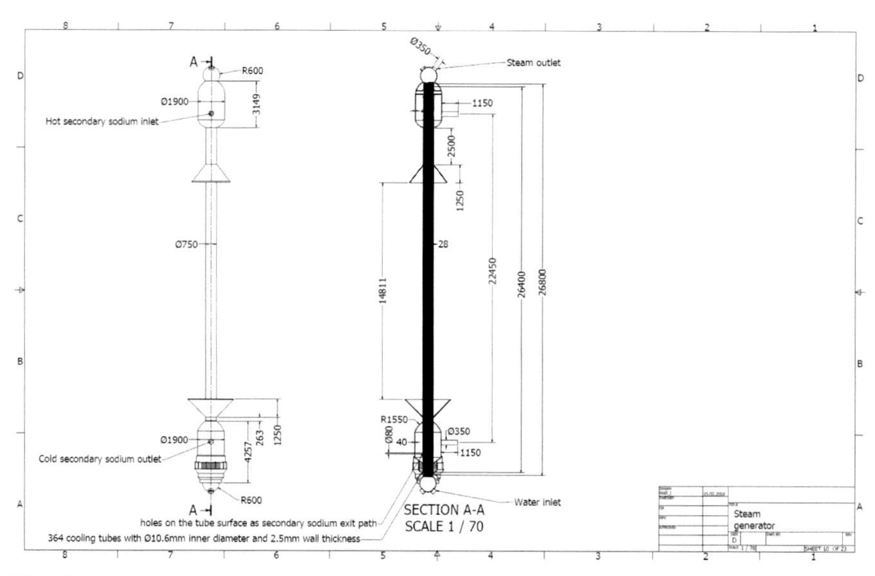

FIG. 13

View of an SG module.

DHRS-1 system

To ensure the highest safety in case of the unavailability of the main heat removal route through the secondary sodium loop, a special DHR system, DHRS-1, is implemented at each of the six IHXs (Fig. 5).

The DHRS-1 loop operates in parallel to the secondary loop using the hot secondary sodium extracted from the IHX as the working medium. The heat is rejected to the environment using a sodium/air heat exchanger located at the bottom of the air stack, which is situated outside of the reactor building. The cold secondary sodium comes back to the IHX cold sodium entry. Such a scheme promotes cooling of the primary sodium in the IHX and therefore enhances the primary sodium natural convection through the core and IHX.

This operation is mainly passive, where the operator has only to open the window of the air circuit and the heat removal starts from the already established sodium circulation (enhanced by the thermal pumps) within the DHRS-1 loop (Fig. 14).

To increase the passivity of the system, a thermal pump of 200 mm diameter can be installed on the hot line. The thermal pump is a passive electromagnetic pump, which uses thermoelectricity generated by the difference in temperatures (hot sodium and atmospheric air). This pump does not need an external electricity supply and provides a supplementary flow rate. The concept of the thermal pump is illustrated in Fig. 15, where a magnetic field is created by permanent magnets. An electric current is produced by the attached thermoelements being exposed to a temperature gradient. The resultant magnetic field and electric current initiate pressure and flow rate in the liquid metal coolant.

The aforementioned component needs further R&D, i.e., testing in a sodium loop. If such R&D is not performed to validate the concept, a small electromagnetic pump with a secured power supply could replace this component.

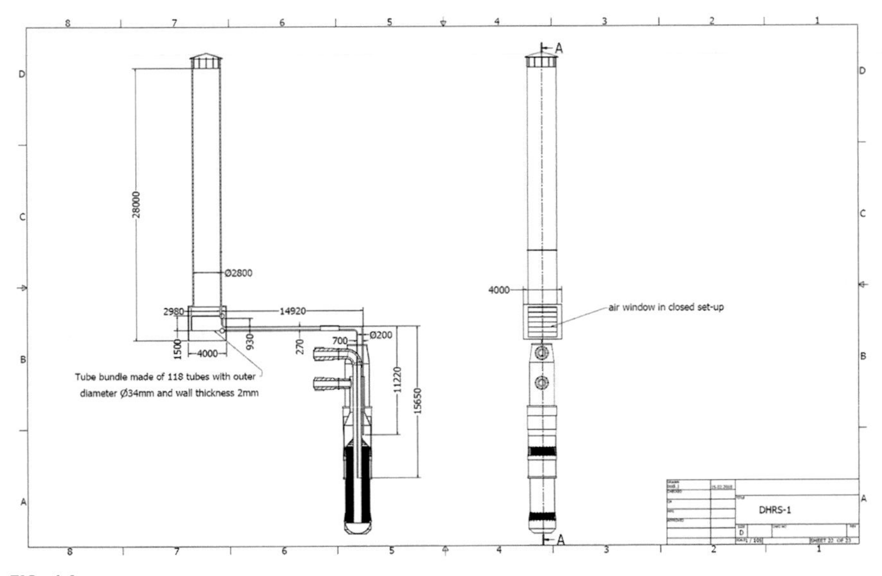

FIG. 14

View of the DHRS-1 system attached to heat exchanger.

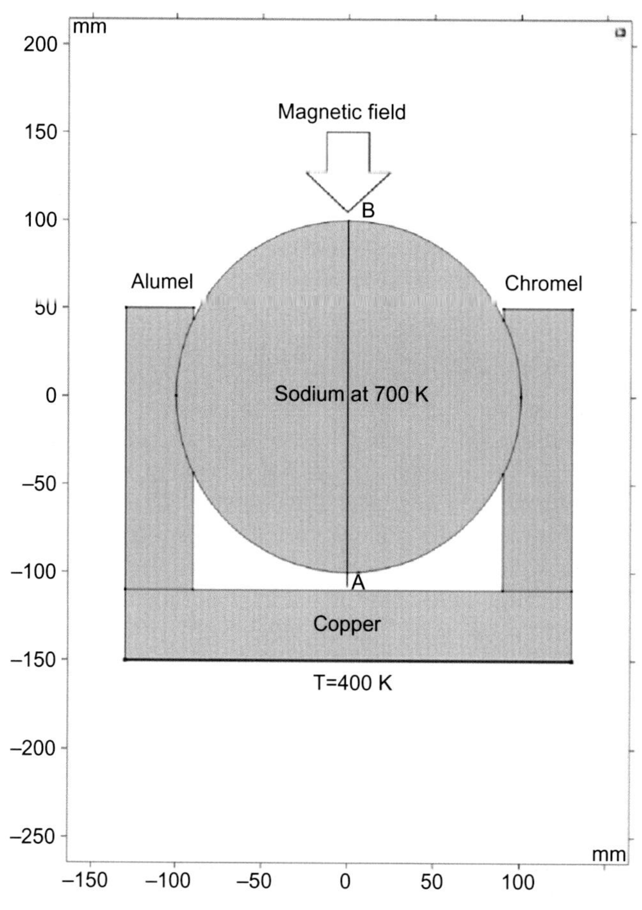

FIG. 15

Principle of thermal pump.

Piping

To limit the sodium velocity to about 5 m/s in normal operation, the proposed diameters of the secondary pipes are Ø 850 mm for the main lines and Ø 350 mm for the SG main lines.

Two options are possible for these pipes:

The first option is to have relatively long and flexible pipes, to be able to accommodate thermal expansion. This has been the option implemented in almost all existing plants and particularly for Phénix and Superphénix. In this case, the main pipes have a wall thickness of 12.5 mm and a length of about 220 m, including elbows to accommodate thermal expansion and discharge lines for sodium draining at the hot and cold legs. SG lines have a wall thickness of 15 mm and a length of about 30 m, including elbows to accommodate thermal expansion and discharge lines for sodium draining at the lower part of each SG. The lengths of the lines depend on the chosen material and its thermal expansion coefficient. The resulting volume of the secondary sodium in each of the six loops is about 250 m^3, including 90 m^3 in SGs, 27 m^3 in IHX, 7 m^3 in the secondary pump, and about 116 m^3 inside the piping.

Indeed, the feedback from these flexible pipes shows some problems. In Superphénix the sodium pipes had lyres to support the dilations between cold and operational states. However, the heavy weight of these pipes requires supports in which the pipes would necessarily need to slide. However, antiseismic standards require pipes to be firmly maintained during an earthquake. This led to the development of rather complex systems on Superphénix that did not work well. After each transient, the pipe was found in abnormal positions.

"The long length of the relatively flexible pipes led to numerous support devices (self-locking devices called DAB) allowing their expansion while blocking them in the event of an earthquake. Many of these numerous devices (2400!) and the nonlinearity of their behaviour made their monitoring and maintenance very cumbersome. Indeed, DAB in bad state could induce blockages of pipes, which would cause significant mechanical stresses." (Guidez and Prele, p. 135)

Another component that worked poorly was the thermal insulation on these flexible pipes. This led to difficulties in detecting leaks, risks of corrosion by undetected leaks, and numerous false alarms that were very difficult to verify [11,17].

On the basis of this negative feedback in terms of investment (significant extra) and safety (risk of rupture of the piping blocked in their support), ESFR-SMART proposes a second option with straight and rigid piping where thermal expansion is taken up by bellows. Fixed and nonsliding supports play their support role in normal operation and in case of an earthquake. Between these fixed points, the pipes are straight. A bellow is installed in the middle of this right part, which supports the dilatation effect. A choice of material other than 316L, such as 9 Cr, would also significantly reduce this dilatation. It should be noted that the Russians on the BN-1200 project chose this bellows option for their design.

The benefits are as follows:

- Cost reduction due to the decrease of the pipe lengths, quantities of secondary sodium, volumes of the storage tanks, etc.
- Simpler, cheaper, and more efficient pipe supports resulting in safety gain and ease to manage in operation.
- Ability to use removable insulation including a gap between this and the sodium pipe, which makes the installation easier on these straight parts and improves the sodium leak detection. This decreases the number of false alarms, which results in improved safety for the reactor operation.
- Improvement of the circuit's natural convection circulation due to the shorter lines.
- Reduction of the distance between fixed points and, thus, the dimensions and cost of secondary building.

- In addition, the mechanical dimensioning of the pipes is simpler than the flexible option, which requires a small thickness for the pipes and numerous welds for the elbows and expansion lyres. With straight piping, it is possible to minimize the number of welds and use pipes with the desired thickness.

In practical terms, on the drawings, straight pipes are used to join the fixed points that are the following components: heat exchangers, SGs, pumps, DHRS-1. In this case, using reduced lengths for the secondary circuits, the implementation of a circular secondary building arrangement is proposed around the primary vessel. This disposition also allows us to use the same chimney for the casing and the DHRS-1. The aforementioned structure of the secondary loop is presented in Fig. 7 (Fig. 16)

The initial flexible loop had a length of 195 m of 850-mm diameter tubing (or 220 m if I count the piping toward the sodium-draining tank) and around 150 m of 350-mm diameter piping. In contrast, with the new design, it was reduced to 67 m of 850-mm diameter tubing (or around 88.5 m if the piping toward the sodium draining tank is counted) and around 12 m of 350-mm diameter tubing. On the basis of the above lengths, the available sodium volume in the piping only (thus, not taking into account the SGs, IHX, or secondary pump) for one secondary circuit for the original case is around 116 m^3, whereas for the simplified circuit this number is around 37 m^3. In this sense, the reduction in sodium volume is 79 m^3.

Safety analysis of the secondary loop

In terms of safety, three points are mainly to be optimized in these secondary circuits:

Firstly, secondary circuits have the role of evacuating the power of the reactor, and in case of a shutdown, it actively participates in the removal of the residual power. The use of these loops has been favored for residual power removal as it is the loop normally used by the operator for this purpose in all operating circumstances. Therefore, we tried to design a loop capable of removing this power by natural convection, passively, and with the minimum of necessary interventions by operators.

Secondly, the operating experience of the SFRs shows that sodium leaks mainly take place at the level of the secondary circuits. For example, in Phénix [8] the 31 leaks of sodium were in the secondary loops and auxiliary systems. Note also that for the ESFR-SMART specific measures have been taken to avoid leakage of primary sodium. These possible leaks of nonactive secondary sodium are more of a security than a safety concern. However, proposals have been made to both minimize the risk of sodium leaks and increase the possibilities of rapid detection and mitigation (see Chapters 3 and 4).

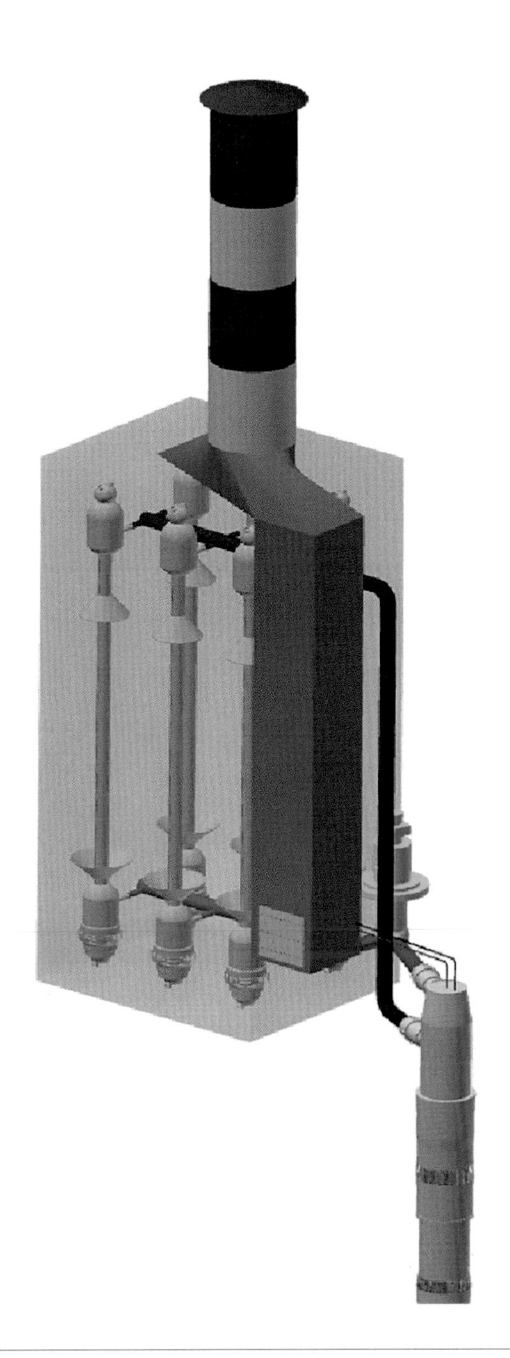

FIG. 16

View of the compact secondary loop with the shared chimney for casing and DHRS-1.

Thirdly, sodium-water interaction is a problem to be tackled at the SG level. Therefore, the SG type was chosen with the aim to improve the speed of detection and minimize the consequences. Corresponding provisions are to be taken at the level of the casings containing the modules.

All these points have been taken into account in the ESFR-SMART design essentially based on the existing feedback experience on the SFR secondary circuits, but also on published results of studies on previous projects such as ASTRID in France, BN-1200 in Russia, or PGSFR in Korea.

- Decay heat removal
  The secondary circuit was sized to be able to evacuate this residual power, even after the loss of the water circuits, only by natural convection of the air around the modules of SGs. However, this sizing revealed that natural convection in the circuits was not sufficient and that an operation of the secondary pumps at reduced speed (100 rpm) was necessary, at least at the start of the event, to increase the heat removal from the circuit [14,16].
  In case of unavailability of the secondary loop and even if this loop is drained, the DHRS-1 can provide 100% of the DHR function, completely passively in natural convection of the air and sodium [14].
  The whole system follows the new Gen-IV safety rules following the Fukushima accident.
- Sodium leaks and fires
  The option of straight lines without elbows allows the use of protection against leakage by a double wall piping, shown in Fig. 17. This installation would be

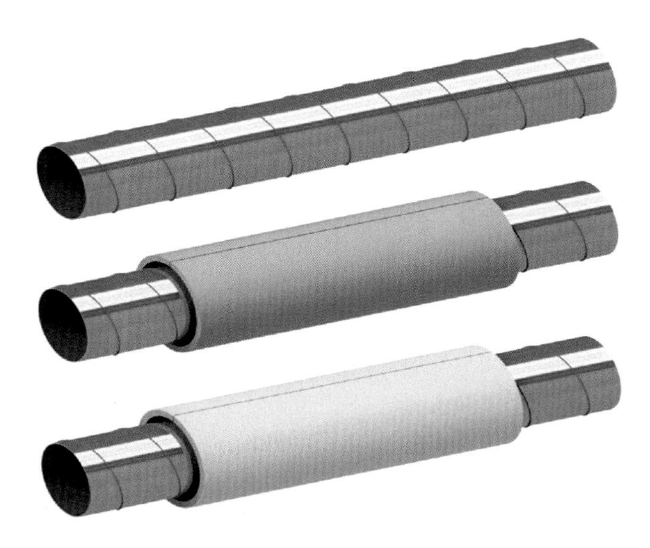

FIG. 17

View of the straight tube with its external and removable insulation.

difficult with large flexible pipes with large movements. The external wall is covered from underneath with insulation, followed by a gap and the sodium pipe. This external wall can be easily opened to allow interventions, for example in case of an alarm.

Classical sodium fire detectors are fitted on the sodium pipe to detect leaks from it. These detectors are installed particularly around the bellows and in the lower part of the circuit. Therefore, sodium leak detection is possible before chemical interaction of the sodium with the insulation. Complementary detectors can be added between the pipe and the removable insulation, such as sodium smoke detectors in the partitioned interior zone. This set of provisions allows quick detection and good containment of sodium leak inside this double wall.

– Sodium-water reaction

Conventional devices enable one to efficiently control the risk of sodium-water reaction by detection of hydrogen in the sodium at the outlet of each SG module. The modularity of the SGs makes it easier to quickly detect a sodium-water reaction and isolate and drain the failed module. It allows also to minimize theoretical envelope accidents. Even with a hypothetical rupture of all the tubes in a module, the accident can be managed in terms of overpressure and mitigation. In the case of sodium-water reaction, the consequences for the plant operations are limited and the operation can continue with the remaining modules, after isolation of the defective module. Mitigation measures against the risk of sodium-water-air reaction must also be considered in the building concept. In particular, "water area" and "sodium area" in the secondary system buildings should be strictly separated to avoid interaction. The casing is sized to resist external aggression.

General layout of the plant

The secondary circuit with short and straight tubes allows a circular disposition of the secondary circuits around the primary block.

This carries an important benefit for the final sizing of the secondary building including all the related pipes and components, as shown in Fig. 18.

This new circular disposition allows significant improvement of the general layout, compared with the initial design with flexible pipes, as shown in Figs. 9 and 10.

We arrive at this final design of the plant with the six chimneys common to SG casings and DHRS-1. On the right we have the turbine building and on the left the handling building with its own chimney (Fig. 21).

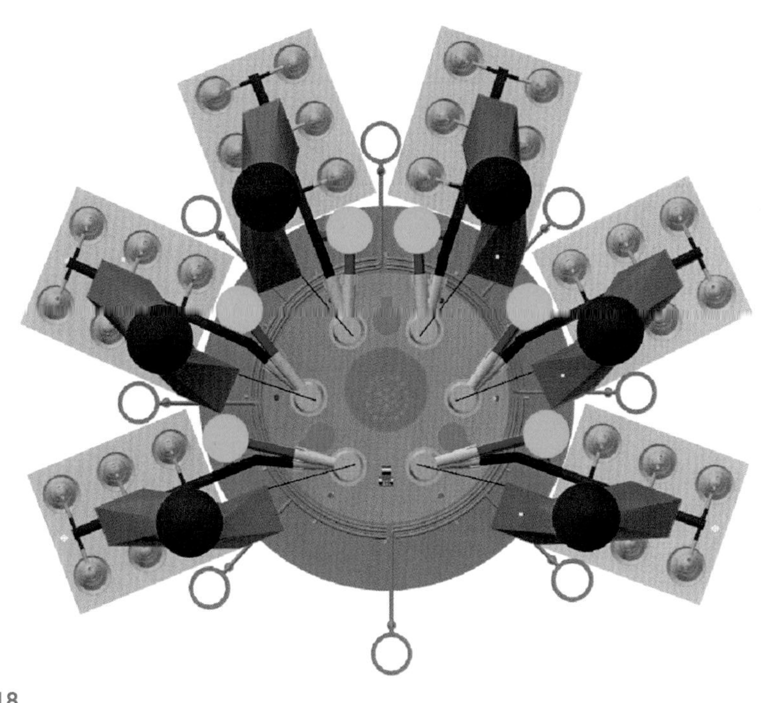

FIG. 18

Circular disposition of the secondary loops around the primary vessel.

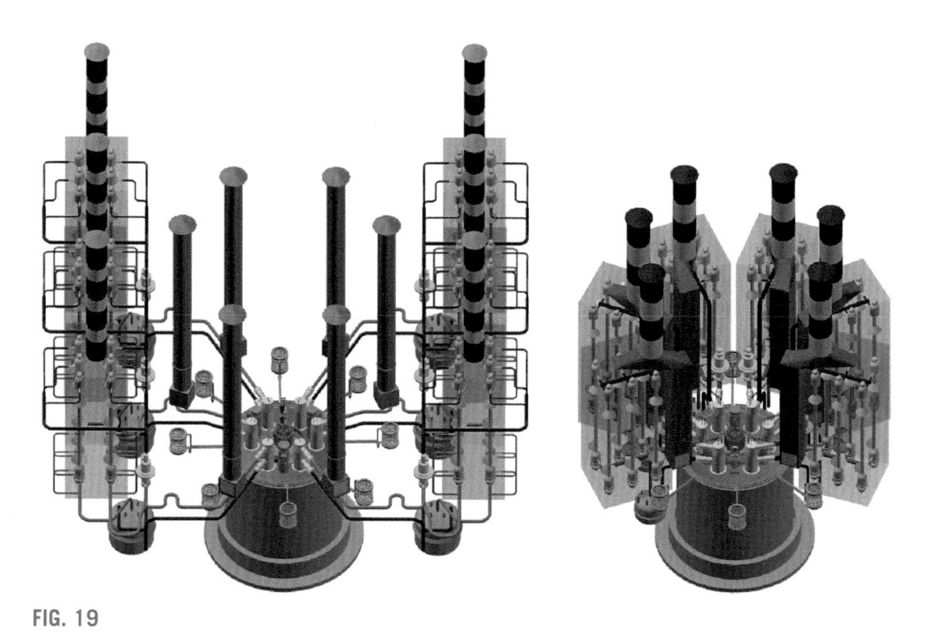

FIG. 19

Original plant layout (left) versus new circular layout (right) in general view of the plant.

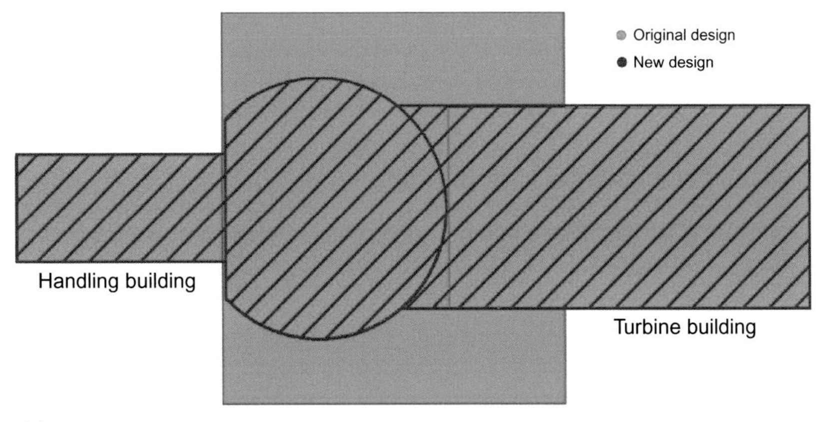

FIG. 20

Comparison of the initial layout and the new circular disposition.

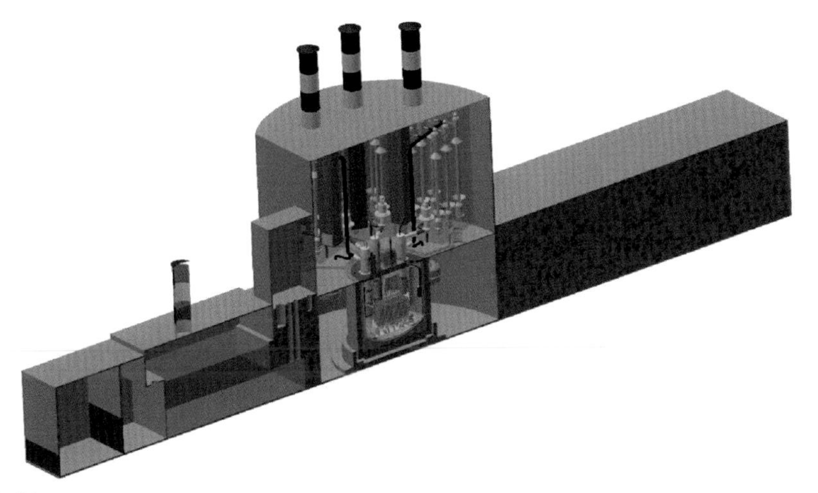

FIG. 21

View of the final layout with a circular layout and with the six common chimneys between the casing and DHRS-1.

Handling systems

Spent fuel handling

Refueling takes place during scheduled reactor shutdowns, which occur annually. Spent fuel subassemblies are removed from the core and placed in the inner spent fuel storage at the periphery of the core. Other irradiated core components, such as reflector or absorber subassemblies, can be removed from the reactor vessel without using this inner storage. Fuel subassemblies that have been retained in the inner spent fuel storage for 3 years and therefore have a low decay heat power (< 10 kW) are transferred from the inner spent fuel storage to the secondary fuel handling facilities.

During reactor operation, special instrumentation systems are continuously used to detect the failed fuel and, once detected, to locate the failed subassembly in the core. This failed subassembly is removed from the core during an exceptional shutdown and placed in a special position in the inner storage, while fresh fuel subassembly is loaded in the core. After reduction of the decay heat power to the level compatible with the secondary fuel handling system, the failed subassembly is removed from the reactor the same way as the other spent fuel subassemblies.

The in-vessel fuel handling system (Fig. 22) provides access to any core position by means of two eccentric rotating plugs (large and small) in the reactor roof, a DLCM, and an FACM. At operational position of the plugs (as shown in Fig. 22), the DLCM is positioned at the center of the ACS. During refueling by rotation of the small and large rotating plugs the DLCM can be positioned above any subassembly from the inner handling zone of the core shown by shaded circle. The subassembly from the inner handling zone can be lifted and moved to the exchange position also shown in Fig. 22, from where it can be taken by the FACM and positioned in the inner spent fuel storage. The FACM is located outside the ACS and by rotation of the large rotating plug can be positioned above any subassembly of the outer handling zone of the core; this subassembly can be moved to the inner spent fuel storage.

The interface with the secondary fuel handling system (Fig. 23) is provided by a two-position rotor suspended from the reactor roof. A fresh fuel subassembly is deposited in the external position of the rotor from the fuel handling cask, while a spent

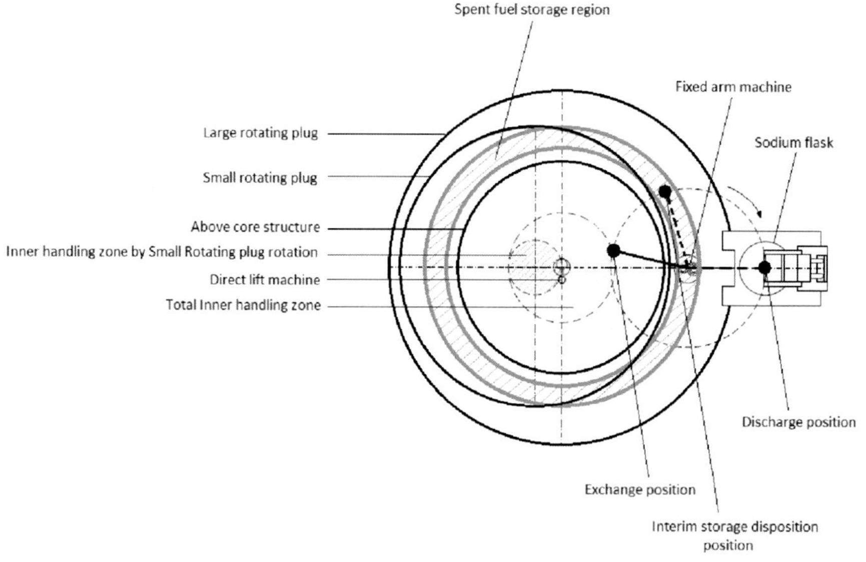

FIG. 22

In-vessel fuel handling system.

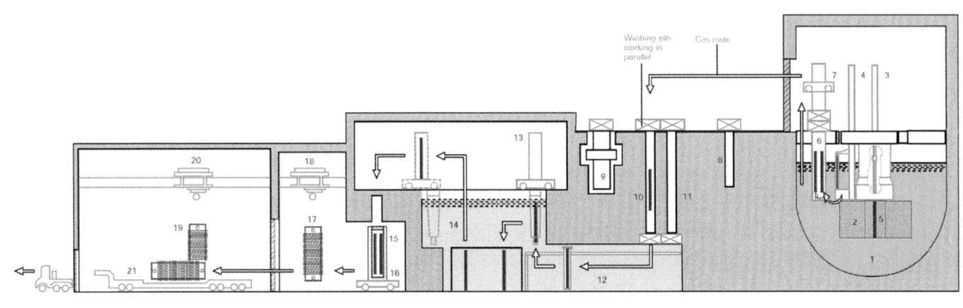

FIG. 23

Fuel handling principle in ESFR-SMART.

assembly is deposited in the internal position by the FACM. After a half turn of the rotor, the fresh fuel subassembly is taken by the FACM and the spent fuel subassembly is lifted into the fuel handling cask. The cask encloses the subassembly in an inert gas atmosphere and incorporates biological shielding. The spent fuel subassembly is transported in the fuel handling cask, along the fuel transfer corridor directly to one of two washing pits in the fuel handling area, where the subassemblies are cleaned. An encapsulation facility is also provided for failed subassemblies. After washing, the spent fuel and breeder assemblies are unloaded through a valve at the lower end of the washing pit, directly into the water transfer corridor through which they are transferred to the spent fuel storage pond. In the case of encapsulation, the subassembly, in its canister, is transferred to a dedicated storage position or to a special transport cask. Absorber assemblies are loaded directly into casks after washing and transported to the reprocessing plant.

After 4-year storage, the decay heat has reduced to about 2 kW. The intact subassemblies are loaded bare into transport casks. These casks leave the fuel handling building via an air lock to an on-site or off-site dry fuel storage for long-term storage. Alternatively, they can be transferred to the reprocessing plant.

Fresh fuel handling

Fresh fuel subassemblies enter the fuel handling building via the transport cask air lock, and are inspected and stored in the fresh fuel dry storage. They are transferred from the fresh fuel storage via the fresh fuel transfer pit into the fuel handling cask and then into the reactor vessel, as described above (Fig. 23).

Handling of components

In case of necessity, in service inspection, repair, or replacement, it is possible to extract the big components, i.e., the IHXs or PPs. The height of the primary building allows these operations to be conducted.

Facilities for washing, decontamination, and maintenance of reactor components are located in the maintenance building, which is separated from the reactor building. Components can be serviced and repaired after washing and decontamination.

Activated and contaminated components are transferred from the reactor building to the maintenance building within shielded transport casks by means of a special-purpose transport system. Within the reactor building, the active component handling system provides the means of removing and replacing both active and inactive reactor components.

The handling system comprises a series of casks sized to accommodate the different components to be handled. In Superphone, a large cask was used for transportation of the big components and an adapted small cask was used for small components or materials, such as cold traps, instrumentation, etc. [4]. The cask encloses the component in an inert atmosphere and incorporates appropriate biological shielding. The casks are transported by the high-integrity overhead crane in the crane hall. They are raised to a fixed height, and a redundant retention feature is engaged that is attached directly to the bridge of the crane. All operations above the reactor are carried out at this fixed height with the redundant retention feature engaged.

Conclusions on safety improvements

The general principle of the studies was to increase the safety in operation, by increasing the simplicity of the design, avoiding adding new systems, and exploiting the possibilities given by the sodium in terms of natural convection and of passivity. We can here summarize the improvements in terms of passivity, simplicity, easy operation, and severe accident mitigation.

In terms of passivity:

- The void reactivity effect is very low, to reduce drastically mechanical energy release in case of accidental sodium boiling. That was obtained by many various dispositions, such as with diameter of pins increased, with a plenum above the fuel assemblies, with mixing of fertile and fissile parts in the core, etc. [11].
- Passive control rods able to stop the plant without control order but only on abnormal variation of a physical parameter such as temperature or flow rate.
- Better design to allow easy natural convection of sodium in the secondary loop, even without water supply and without AC power supply.
- Possibility of power removal without water supply, only by natural air convection, in the casing containing the six modules of SG.
- A passive DHR system on each loop able to maintain a cold leg in the heat exchanger by passive way with air, even if the secondary loop is drained.
- Thermal pumps, passive, able to maintain permanent flow rates in the secondary loops and in the DHRS-1, even without AC power supply.

In terms of simplicity:

- Suppression of the safety vessel.
- No dome or polar table.

- No DHRS inside the primary vessel.
- Minimization of the number of sodium circuits.
- Very simple massive roof.
- Reduction of more than 50% of pipe length and of general reactor layout, with the use of straight pipes in secondary loops.

In terms of operation:

- New measures against sodium leaks and better protection of the building with strong separation of water and sodium circulation areas.
- Better concept to avoid primary sodium leakage.
- Better access for handling operations (no polar table).
- Quick water sodium reaction detection and good protection against consequences based on choice of modular SG.
- Dispositions to avoid, by design, gas entrainment in the core.
- Reactor very forgiving with a high inertial capacity and possibility to wait a long time without operator action.
- Minimization of the number of sodium circuits to operate and survey.

In terms of severe accident mitigation:

- Discharge tubes inside the core to stream the corium to the core catcher in mitigation situation (Fig. 24).
- Low potential for mechanical energy release with a new core concept.
- In case of severe accident, big mechanical margins with the massive reactor roof and with the reactor pit able to withstand sodium leaks. This ensures that there is no radioactive release in the short and long term.

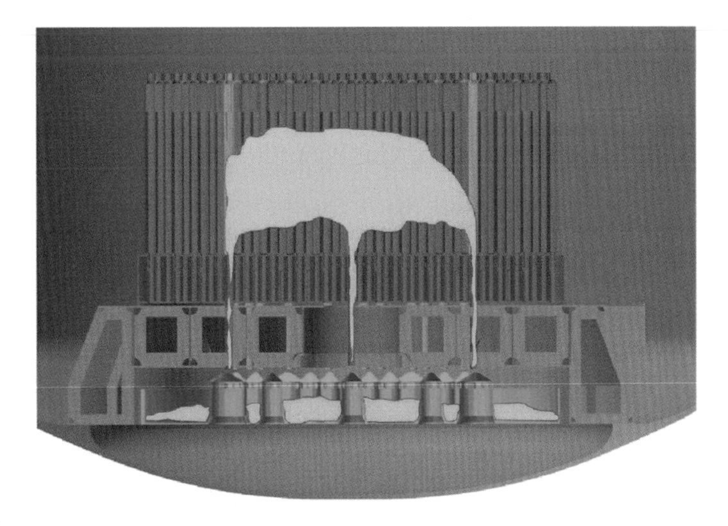

FIG. 24

"Artistic view" of the preferential ways for the melted core.

– Ability to cool the primary vessel for long duration after severe accident with three independent systems, each being sufficient alone.
– A dedicated core catcher able to receive the whole core materials, with materials protecting the core catcher against ablation by corium, with efficient natural convection cooling and without recriticality potential.

The proposed set of modifications, compared with the EFR and CP ESFR design, aims at consistency with the new safety rules for Generation-IV SFRs since the Fukushima accident.

R&D needs for ESFR-smart options

To evaluate the options for the ESFR-SMART design, calculations were made to provide a preliminary assessment and to assess the feasibility of various options:

– The pit organization is presented in Ref. [12], with thermal calculations in nominal and accidental situations. The thermal possibilities of the DHRS-3 were also calculated.
– Thermal calculation of the three DHR systems was made. For the DHRS-2, Cathare calculations were necessary to calculate the natural convection and the cooling by air in the SG casings. These calculations show the compliance of the three DHR systems with the new Gen-IV safety rules [14,15].
– For the core catcher, calculations are provided for core melting, residual power, and the ability of the core catcher to evacuate by natural convection of sodium around this power [16].
– Some calculations are provided on the thermal pumps for presizing of DHRS-1.

All these calculations, to verify the feasibility of the options, are explained in the ASME article in Ref. [18].

However, especially if one day Europe wanted to build a reactor on these bases, some points remain to be addressed. They require little R&D and would relate mainly to the following points:

– **Industrial confirmation of the proposed organization for the reactor pit**
  The organization proposed is based on developments already made for EFR, such as testing of an aluminous concrete without reaction with sodium. The global organization of the pit should be validated.
– **Industrial validation of the manufacturing method of the EFR-type thick slab**
  Thick slab needs thick welds. This type of operation has already been manufactured. However, the global organization of the slab fabrication, initial part in factory and final welding on site, must be managed industrially.
– **Qualification of low-expansion materials and large-diameter bellows for the secondary circuit**

FIG. 25

View of the bellow tested on Superphénix heat exchanger [4].

Further R&D is necessary for the bellows of 850-mm diameter in terms of expansion capacity and operating lifetime. However, the use of bellows on sodium loops is not unprecedented.

These bellows exist on many sodium valves, especially in Phénix and Superphénix, and inside the Phénix heat exchangers.

A bellow of large diameter (~ 800 mm) was installed in Superphénix (Fig. 25) on the internal part of the hot collector of the IHXs to take up the differential expansions with the external part. This device had several expansion waves and a thickness of 8 mm. It has undergone a cycling test with a large number of cycles for validation.

On the ESFR-SMART SG module, a bellow of large diameter (750 mm) is designed to allow relative dilatation between the external wall of the SG and its internal bundle.

Depending on the dimensions of the circuit, this R&D consists of defining the specifications requested for these bellows in terms of resumption of expansion. Then it is necessary to build some industrially, with the ability to execute a large number of cycles in a furnace, as was done for the Superphénix bellows.

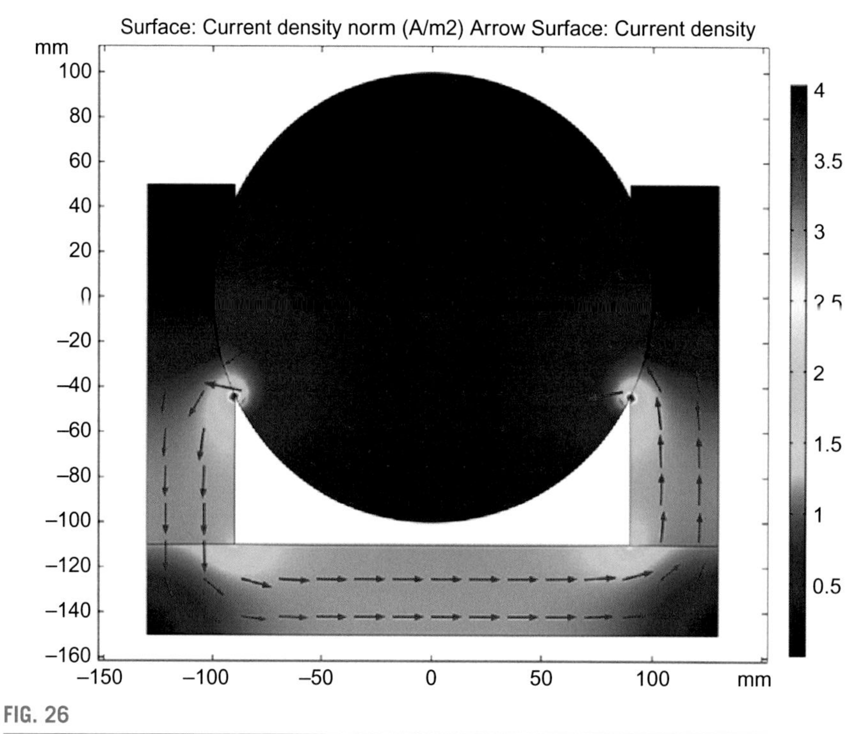

FIG. 26

Computed current density in DHRS-1 thermal pump (A/m^2).

Another aspect of R&D is to study and test the pipes and new materials with lower expansion coefficients. Some are already available but need to undergo testing to confirm their effectiveness.

- **R&D on thermal pumps**

  Thermal pumps are not new and were already used on the Siloe reactor to ensure the flow in the test loops. However, although some calculations were performed on the thermal pump of the DHRS-1, a full-scale test on a sodium loop would be necessary for final validation and industrial demonstration of the results (Fig. 26).

Conclusion

Thanks to the ESFR-SMART project, precalculations and a set of new safety measures are already available for the design of a SFR, meeting the new reinforced post-Fukushima safety criteria. This reactor brings about significant simplifications, incorporating feedback from previous European reactors and projects. These simplifications bring safety improvements, cost savings, and ease of operation.

All these improvements are also available and useful for design of SFR with lower power.

This project makes it possible to have a database available for European development of these reactors in the future, which could potentially solve the energy problems of humanity. However, today, there are no short-term projects in Europe.

Acknowledgment

The work was prepared within EU Project ESFR-SMART, which received funding from the EURATOM Research and Training Programme 2014–18 under grant agreement no. 754501.

References

[1] EFR Associates, European Fast Reactor: Outcome of Design Studies, 1998.

[2] G.L. Fiorini, A. Vasile, European Commission—7th Framework Programme. The Collaborative Project on European Sodium Fast Reactor (CP ESFR), Nucl. Eng. Des. 241 (9) (2011) 3461–3469, https://doi.org/10.1016/j.nucengdes.2011.01.052.

[3] K. Mikityuk, E. Girardi, J. Krepel, E. Bubelis, E. Fridman, A. Rineiski, N. Girault, F. Payot, L. Buligins, G. Gerbeth, N. Chauvin, C. Latge, J.C. Garnier, ESFR-SMART: new Horizon-2020 project on SFR safety, in: Proceedings of International Conference on Fast Reactors and Related Fuel Cycles: Next Generation Nuclear Systems for Sustainable, Development (FR17), Yekaterinburg, Russia, 26-29 June, IAEA-CN245-450, 17 pages, 2017.

[4] J. Guidez, G. Prêle, Superphénix. Technical and Scientific Achievements, Edition Springer, 2017, ISBN: 978-94-6239-245-8. 342 p.

[5] The Generation IV International Forum: The Safety Design Criteria Task Force (SDC-TF), Safety Design Criteria for Generation IV Sodium-cooled Fast Reactor System (Rev. 1), 2017, SDC-TF/2017/02, 91 pages, Free download from: https://www.gen-4.org/gif/upload/docs/application/pdf/2018-06/gif_sdc_report_rev1-sept30-2017_aftereg-pg20171024a.pdf.

[6] WENRA, Safety of new NPP designs, 2013, WENRA/RHWG Report, 52 p. Free download from: http://www.wenra.org/media/filer_public/2013/08/23/rhwg_safety_of_new_npp_designs.pdf.

[7] IAEA Nuclear Energy Series No. NR-T-1.16, Passive Shutdown Systems for Fast Neutron Reactors, Vienna, Austria, 2020, Free download from: https://www-pub.iaea.org/MTCD/Publications/PDF/P1863E_web.pdf.

[8] J. Guidez, Phénix Experience Feedback, 2014. Edition EDP Sciences, EAN13: 9791092041057, 300 p.

[9] IAEA-TECDOC-1691, Status of Fast Reactor Research and Technology Development, IAEA TECDOC Series, Vienna, Austria, 2012, 846 p. Free download from: https://www-pub.iaea.org/MTCD/Publications/PDF/TE_1691_CD/PDF/IAEA-TECDOC-1691.pdf.

[10] R. Baldev, P. Chellapandi, P.R. Vasudeva Rao, Sodium Fast Reactors with Closed Fuel Cycle, CRC Press, 2017. November 15, ISBN 9781138893047, 720 p.

[11] A. Rineiski, C. Mériot, M. Marchetti, J. Krepel, C. Coquelet-Pascal, H. Tsige-Tamirat, F. Alvarez-Velarde, E. Girardi, K. Mikityuk, New ESFR-SMART core safety measures and their preliminary assessment, submitted to ASME, J. Nucl. Eng. Rad. Sci. (2021). 26 p.
[12] J. Guidez, A. Gerschenfeld, E. Girardi, K. Mikityuk, J. Bodi, A. Grah, European Sodium Fast Reactor: innovative design of reactor pit aiming at suppression of safety vessel, in: International Congress on Advances in Nuclear Power Plants (ICAPP 19), Juan-les-Pins, France, May 12-15, 10 p, 2019.
[13] V.Y. Sedakov, M.A. Lyubimov, V.I. Shkarin, V.P. Shokhonov, Manufacture, Installation and Adjustment of BN-800 RP Equipment, in: Proceedings of International Conference on Fast Reactors and Related Fuel Cycles: Next Generation Nuclear Systems for Sustainable, Development (FR17), Yekaterinburg, Russia, 26-29 June, IAEA-CN245-425, 10 p, 2017.
[14] J. Bittan, C. Boré, J. Guidez, Preliminary assessment of decay heat removal systems in the ESFR-SMART design: the role of natural air convection around steam generator outer shells in accidental conditions, in: International Youth Nuclear Congress (IYNC), March 8-12, Sydney, Australia, 2020.
[15] J. Guidez, A. Gershenfeld, K. Mikitiuk, J. Bodi, E. Girardi, J. Bittan, C. Bore, A. Grah, Innovative decay heat removal systems in European Sodium Fast Reactor, in: 2020 International Congress on Advances in Nuclear Power Plants (ICAPP, 2020.
[16] J. Guidez, A. Gerschenfeld, J. Bodi, K. Mikityuk, F.A. Velarde, P. Romojaro, U. Diaz-Chiron, ESFR Smart Project Conceptual Design of in-vessel Core Catcher, in: Physor, 2020.
[17] J. Guidez, A. Saturnin, Evolution of the collective radiation dose of nuclear reactors from the 2nd through to the 3rd generation and Generation-IV sodium-cooled fast reactors, in: Proceedings of International Conference on Fast Reactors and Related Fuel Cycles: Next Generation Nuclear Systems for Sustainable, Development (FR17), Yekaterinburg, Russia, 26-29 June, IAEA-CN245-016, 12 p, 2017.
[18] J. Guidez, E. Girardi, K. Mikitiuk, J. Bodi, B. Carluec, A. Grah, New reactor safety measures for the European Sodium Fast Reactor. Part II: preliminary assessment, 2020. submitted to ASME J. Nucl. Eng. Rad. Sci., NERS-20-1080.

CHAPTER

Update on molten salt fast reactors

7

Abstract

Among the fourth-generation reactors, the concept of molten salt reactors is particularly original. In this concept, the nuclear fuel is melted in a salt, and the nuclear reactions take place in this mixture, which then transmits the energy released to an intermediate fluid. It is therefore the only type of reactor without solid fuel assemblies within a coolant.

This original design has many potential advantages:

- In terms of safety, there are no longer any serious accidents with fuel-coolant interaction. As the salt circuits are not under pressure, these reactors have intrinsic safety, improving their social acceptability.
- The fuel cycle is simpler and faster, and generates less waste since we no longer have to cut assemblies before reprocessing.
- The incineration of all the minor actinides is facilitated, which makes it possible to generate only short-lived waste that is more easily manageable.
- The salt + fuel volume is much more compact than the solid fuel + heat transfer fluid assemblies.
- The piloting done by power call to the intermediate circuit allows very reactive network monitoring.

It is for this reason that many projects are currently under study in China, Russia, the United States, etc., with demonstration reactors under construction. Many types of reactors are possible depending on the choice of salts and the choice of fissile or fertile, in thermal or fast versions.

Unfortunately, experience feedback is weak for this type of reactor. Only one American experimental reactor, the MSRE, operated successfully, in the 1960s, as a thermal reactor cooled with salt fluoride.

Because of this limited experience feedback, many problems need to be studied and validated, including choice of salt, choice of materials, choice of fuel cycle, neutron/thermohydraulic coupling, development of computer codes, operating mode, and salt reprocessing.

In summary, these MSR reactors in their fast version have great potential but require developments to be validated on demonstration reactors.

This chapter reviews the state of knowledge and developments in this field.

Fast Reactors. https://doi.org/10.1016/B978-0-12-821946-1.00007-3

Keywords

Molten salt, Plutonium, ARE, MSRE, Fluoride, Chloride, Minor actinides, Fuel cycle, Incineration, MSFR

Principle of operation of molten salt reactors

Unlike all other reactors, which have solid fuel cooled by a coolant, the principle of molten salt reactors is to dissolve nuclear fuel in molten salt at high temperature. When a certain critical mass is reached during the filling of the tank, there is divergence and production of a power, which leads to the heating of the fluid and its expansion. This dilation decreases the reactivity with negative feedback coefficients, and a new equilibrium temperature is reached. Adjusting the fuel additions makes it possible to reach the desired equilibrium temperature for the fuel salt circuit. This combustible salt will then remain at this new temperature without producing power.

For the extraction of power, it is necessary to put this salt in circulation through exchangers coupled to an intermediate circuit, itself coupled to a classic tertiary circuit, for example, in water. The start of the cooling by the intermediate circuit will return to the inlet of the tank a cooled combustible salt, which triggers divergence and production of power. Therefore, the power demand by the intermediary controls the reactor.

The operation of the reactor requires regular additions of fresh fuel, and partial reprocessing to gradually remove certain products created that are neutron poisons. It will also be necessary to manage the fission gases produced by nuclear reactions. We arrive at the basic diagram below (Fig. 1).

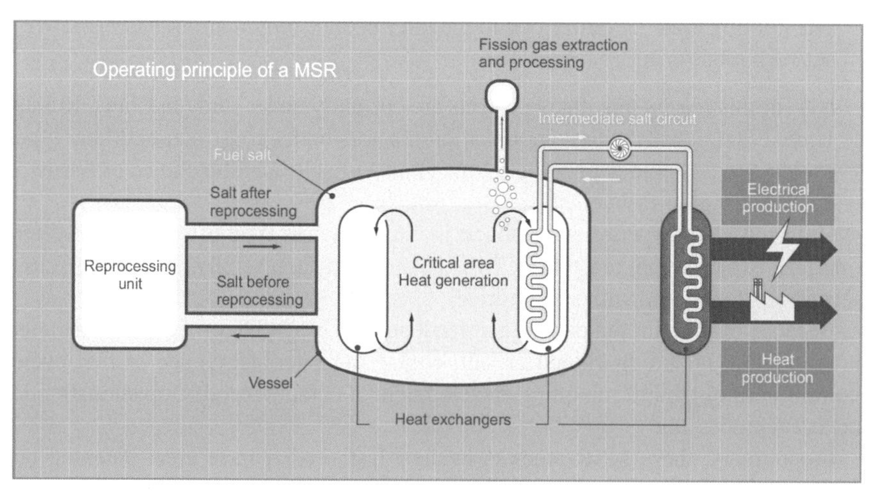

FIG. 1

Schematic diagram of operation of an MSR.

We immediately see that the concept has significant potential advantages, which will be detailed later in terms of the fuel cycle, safety, compactness, network monitoring, etc.

The possible types of MSR

Molten salts can be used as a heat transfer fluid in certain concepts with solid fuel, in particular with the fuel of high-temperature reactors (HTRs). The aim is to replace helium in these HTRs with molten salt, which has potential advantages. We will not discuss this concept here, as it does not fit into our framework of studies devoted to fast flow reactors. However, it should be noted that these concepts lead to significant technological developments in the use of molten salt, for example, on exchangers, instrumentation, and pumps (see, e.g., the Massachusetts Institute of Technology (MIT) report in Ref. [1]. These developments are, of course, also very useful for MSRs.

Different types of MSR exist since there are several elements of choice in terms of salts, and fissile or fertile fuels.

At the level of the fissile, the choice remains limited. In the first order, enriched ^{235}U is the most accessible, then plutonium for countries that do reprocessing. With advanced reprocessing, it is also possible to have americium-type minor actinides. It should be noted that the use of thorium as fertile makes it possible, after reprocessing, to have a new high-performance fissile: ^{233}U.

At the fertile level, there are, of course, stocks of either natural or depleted uranium, and there are significant amounts of thorium available.

A wide range of salts can be used. However, as the choice is more complex, we devote Appendix 3 of this book to it.

Several factors come into play, in particular:

- An acceptable melting temperature (often obtained by creating eutectics).
- Correct product solubility in the desired operating range.
- Good neutron behavior.
- Possibilities of reprocessing.
- Feedback.

We see that, by combining the choice of fissile products, the choice of fertile products, and the choice of the salt (or the eutectic) carrying the assembly, we arrive at a large number of possibilities.

To simplify (see developments in Appendix 4), we arrive at two main families: fluoride salts and chloride salts. Each of these salts has advantages and disadvantages, but it should be noted that thorium and uranium have good solubility coefficients in fluoride salts and plutonium has better solubility in chloride salts and insufficient solubility in fluoride salts.

Projects with fluoride salts use enriched uranium as fissile and thorium as fertile, with the aim of creating ^{233}U, extracting it during reprocessing, and developing a medium-term thorium sector/^{233}U.

Projects with chloride salts use plutonium as a fissile, or other available minor actinides such as americium, and depleted uranium or even thorium as fertile.

It should be noted that the solubility of minor actinides as americium, is sufficient in fluoride salts, which is exploited in the Russian MOSART project.

All these reactors can be designed in burner mode if we want to eliminate stocks of plutonium or americium. To achieve this, it is enough to simply not add fertile material, which will lead these reactors to consume the fissile products.

To design a reactor in isogenerator modes, it is enough to add an appropriate amount of fertile, which will create fissile products. For example, depleted uranium will create plutonium and compensate for its disappearance. The reactor is then a fertile burner.

Finally, to design a fast breeder reactor, add fertile blankets around the reactor, which will produce additional fissile products. These blankets can consist of depleted uranium or thorium. The reactor is then a fertile burner and a fissile producer.

Experience feedback on MSRs

The MSR concept dates back to the 1950s, when this concept of a reactor with fuel dissolved in molten salt was developed in the United States for use on a military aircraft with a very long range of action (we are in the middle of the Cold War!). The choice of this type of reactor came from its intrinsic potential for very fast load following, necessary for an aircraft.

A prototype, the Aircraft Reactor Experience (ARE), was in operation only from November 3 to 12, 1954. Although its design had some strengths, problems of corrosion of the materials used (Inconel) and the complexity of the installation brought the military career of the MSR to a halt, based on this experience acquired at the MSRE. This end of its military career was due to other reasons as well, including the dissemination of radioactive products in the event of a plane crash.

This higher-power experimental reactor (7.4 MWt) operated from 1965 to 1970. In 1976, on the basis of this experience, a project for a higher-power-generating reactor, the Molten Salt Breeder Reactor (MSBR), was designed and then abandoned.

Since then, no reactor, not even a prototype, has been built, thus limiting the operational feedback to the feedback from the MSRE prototype.

The MSRE is a thermal spectrum reactor using graphite moderator and fluoride salts, with a power of 7.4 MWt. It was initially loaded with 33% highly enriched uranium.

The design studies started in 1960, and the construction ended in 1964, followed by a divergence in 1965. It operated from 1965 to 1969. There is extensive open-access documentation on this operation (several thousand pages; see refs. [2,3]).

This reactor was a loop reactor (Fig. 2), unlike most of the current projects, which concern integrated reactors.

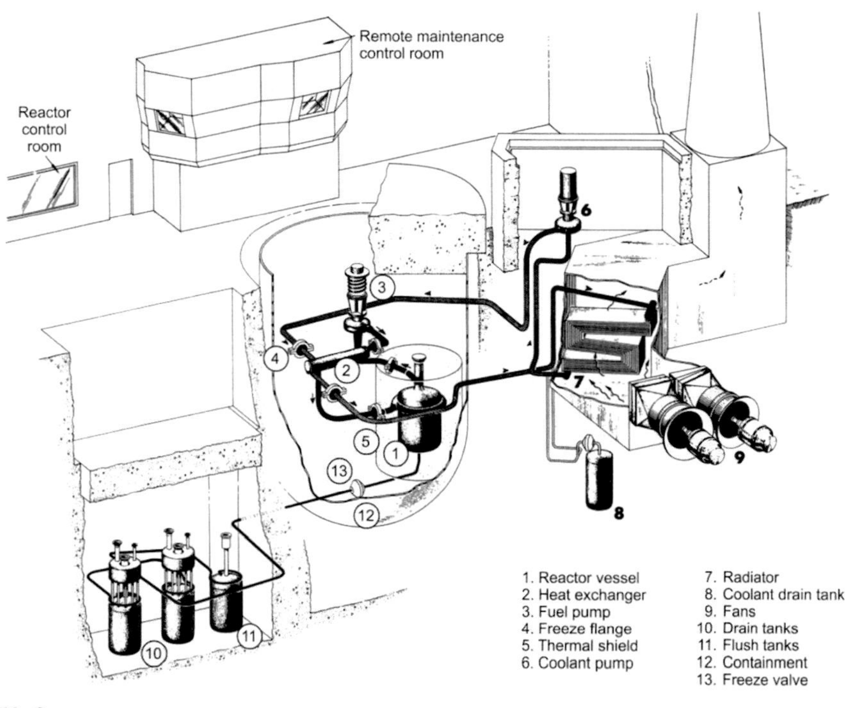

FIG. 2

General view of MSRE.

Choice of salt (FLiBe)

On the basis of the ARE experience, lithium and beryllium fluorides (FLiBe) were chosen (Fig. 3), which have the following characteristics:

- Good neutron economy (lower capture section).
- Beryllium lowers the melting temperature.
- The use of Li-7 prevents the formation of tritium.
- The addition of ZrF_4 to prevent the formation of uranium oxides was mentioned but did not prove necessary.
- The good behavior under irradiation has been confirmed.

The materials

The corrosion of Inconel on the ARE led to the development of a material: Hastelloy N, a nickel-based material with a high molybdenum content. This material has shown excellent resistance to corrosion by salts.

However, shallow embrittlement by tellurium was observed. In fact, it is the fission products that are potential corrosion factors, requiring control of the redox potentials.

FIG. 3

Molten salt.

Neutronics and flows in the primary

The "cold" salt (635°C) descends at the periphery to keep the tank "cold," then in the lower plenum shifts to an upward flow in graphite channels, which are the moderator, where a temperature of 665°C is reached.

The graphite channels keep the reactor in the thermal spectrum, and the low power causes a low core inlet-outlet temperature difference.

There are 513 graphite channels, occupying 77.5% of the volume.

The flows are laminar (Figs. 4 and 5).

Three control bars were planned, held by electromagnets and without contact with the salt. Each bar had 38 gadolinium-based cylinders. These bars had some technical difficulties and were not very useful.

Draining devices were provided and used in maintenance procedures.

A primary pump in the upper part ensures the circulation of the primary fluid in an expansion tank, where the extraction of fission gases and any additions or withdrawals take place (Fig. 6). Sealing at the level of the passage of the pump shaft is ensured by a descending current of helium, which then entrains the fission gases for subsequent separation.

The main problems encountered included aerosol deposition, plugging of filtration systems, and fluctuating gas entrainment in the salt. This led to power fluctuations, which were very sensitive to the gas content.

Heat extraction

This extraction was carried out at the primary level by exchangers (total diameter 40 cm, made up of 163 tubes 2.5 m long) and at the secondary level by a radiator of 120 tubes 9 m long with a fan (Fig. 7).

The only problem was a fan failure for 11 weeks.

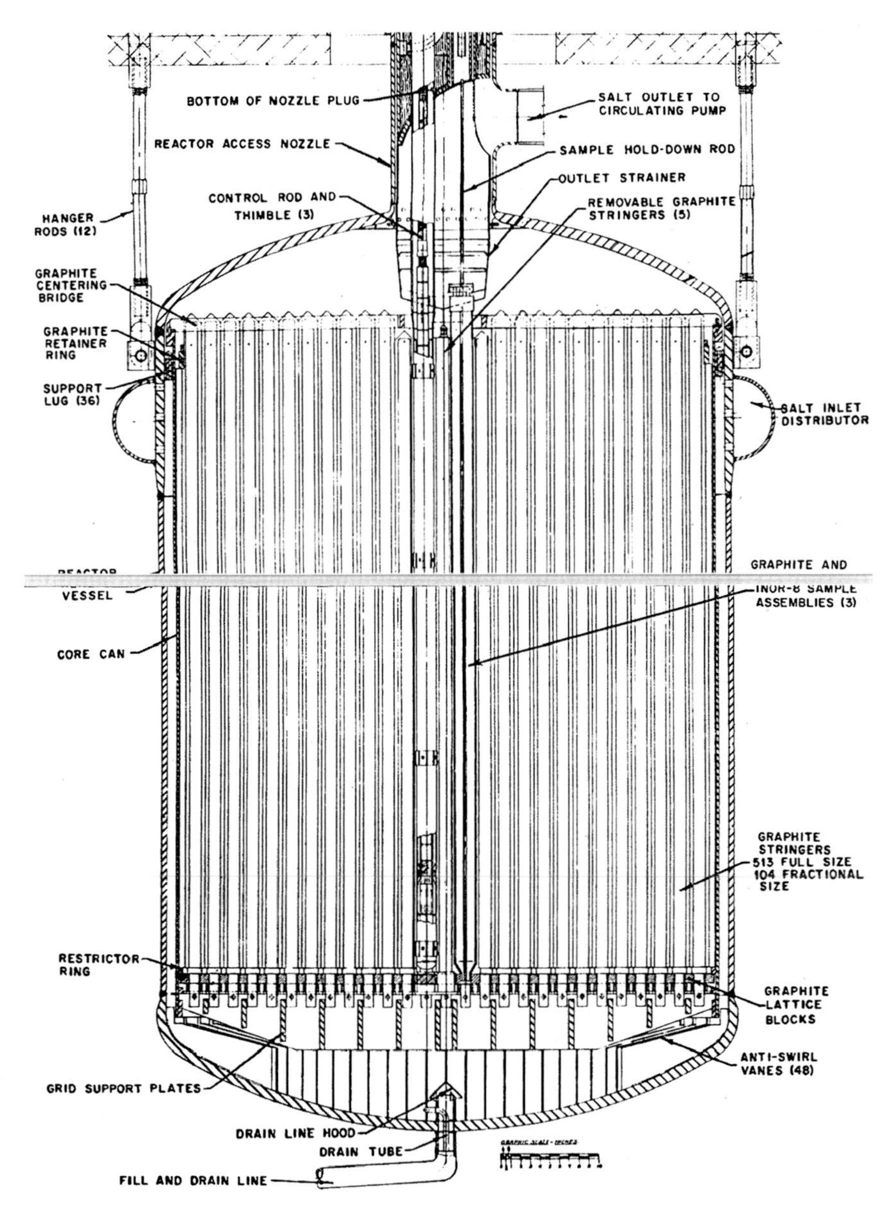

FIG. 4

View of the core of the MSRE reactor.

FIG. 5

View of the MSRE core graphite element.

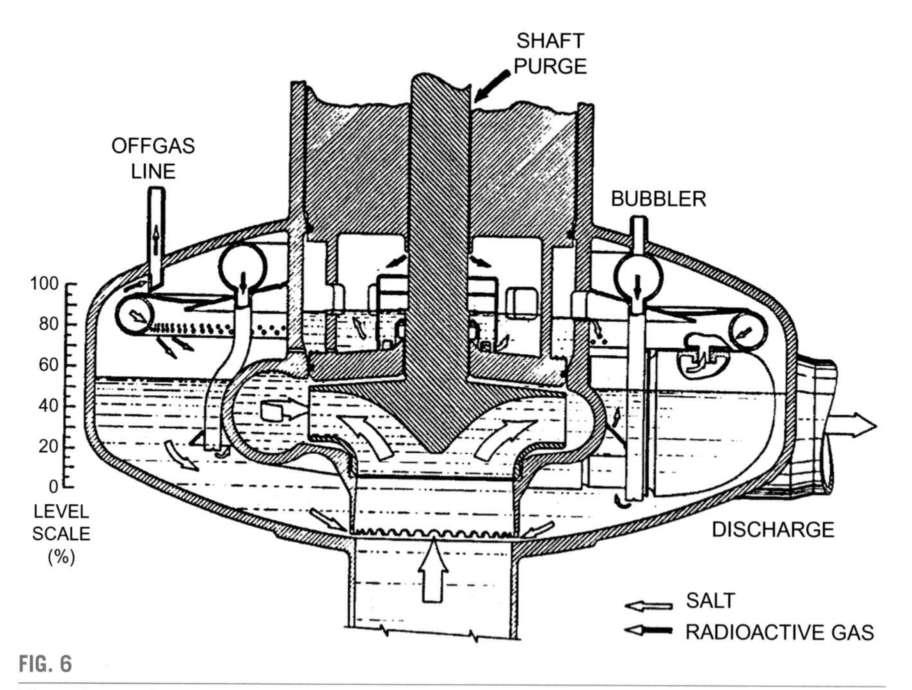

FIG. 6

View of the primary pump of the MSRE.

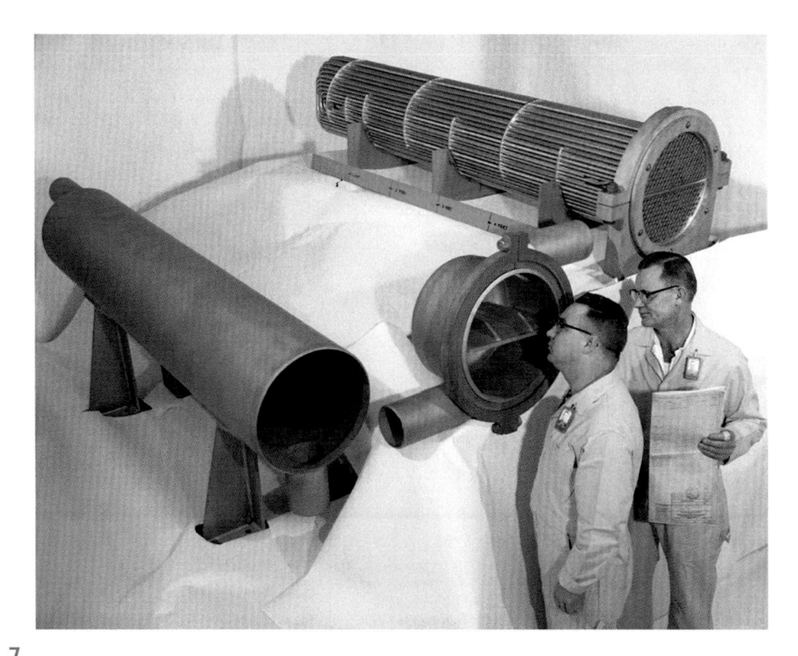

FIG. 7

View of MSRE exchangers.

Salt processing

The gaseous fission products (mainly Kr and Xe) are extracted by injecting helium into the expansion tank of the pump.

Redox control is ensured by guaranteeing the presence of UF_3 to limit the excess of F caused by fissions. In practice, this is achieved by immersing metallic beryllium in the pump vessel in order to form BeF_2 and thus reduce part of the UF_4 to UF_3.

When switching from ^{235}U to ^{233}U, fluoridation of the salt made it possible to extract the ^{235}U from the primary salt. The added ^{233}U came from the reprocessing of assemblies with thorium irradiated in various American pressurized water reactors (PWRs) (Indian Point in particular). The separation of this ^{233}U was carried out at the Thorium-Uranium Recycle Facility in Oak Ridge.

After the reactor was shut down in 1969, the salt remained in on-site storage tanks. In 2005, the extraction of salt from ^{233}U began (Refs. [2,3], and Fig. 8). The operation ended in 2008.

Fission products have never been extracted from the salts, which are still stored at the site.

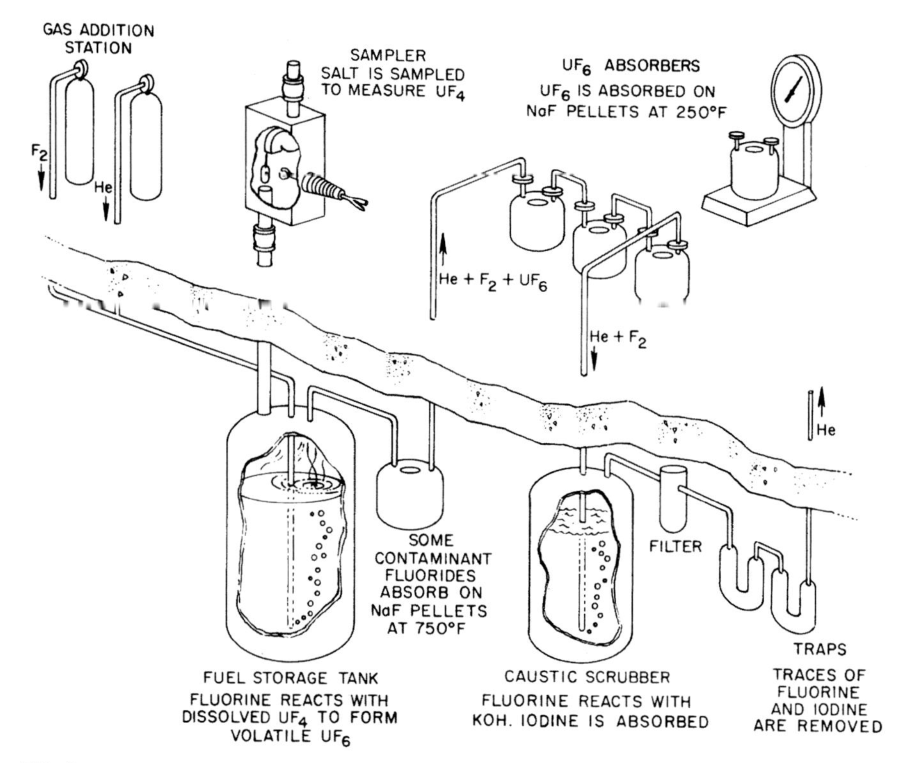

FIG. 8

General view of ^{233}U recovery operations.

Final balance sheet

The reactor operated for 15 months, 6 of which were continuous without draining the salt, with two major campaigns: one where ^{235}U enriched to 33% was the fissile fuel, and one where ^{233}U played this role.

Contrary to popular belief, thorium has never been used as fertile. This role was performed by the initial uranium-238. Therefore, plutonium was created, but the quantities remained low given the short duration of operation (estimated at 600 g).

The equivalent time at full power was 9005 h in ^{235}U and 4167 h in ^{236}U (13,172 h in total) for a total fuel circulation time of 21,788 h and a total secondary circulation time of 26,026 h.

There were 51 primary fill/drain cycles and 19 secondary fill/drain cycles.

Over this relatively short period of time, the operation went well with a small series of minor problems to be solved: rupture of the fan, gas entrainment, failure of bars to fall, cracking of a valve during emptying, etc.

As the main poison, xenon, was recovered in gaseous form continuously, there was no need to reprocess the salt to be able to operate over this period. Only uranium-233 was extracted and recovered during a subsequent campaign. The salt itself is always stored with all the fission products on site. There was therefore no reprocessing and no closing of the cycle.

After the abandonment of the MSBR project, the Idaho National Laboratory transcribed its experience into fully open and accessible quality documents, which today remain a frequently used database.

The potential benefits of MSRs

The concept of molten salt reactors has some potential advantages:

- A criticality incident does not lead to core meltdown and interaction phenomena. So, the energy releases are much lower, which facilitates the analysis of the safety of the containment.
- The fuel cycle is much simpler since there is no longer a need to manufacture fuel. It suffices to supply fuel to the core by periodic addition.
- Some concepts make it possible to burn all the waste produced by PWRs and to minimize the final waste of all nuclear activities.
- The fast versions have all the advantages already mentioned in Chapter 2.
- The assembly presents great virtual compactness.
- Network tracking is very easy and efficient.

These potential advantages are now leading to a great deal of research around the world on this concept, either within the framework of national programs or within the framework of startups, attracted by the originality of the concept and its potential advantages.

The advantages of MSR in terms of safety

MSRs are the only reactor concept where there is no separation between the coolant and the solid fuel. However, serious accidents on a "conventional" reactor are due to situations of interaction between the coolant and the fuel (see Three Mile Island, Chernobyl, and Fukushima). Indeed, the interaction between the fluid and the molten fuel leads to significant energy releases. It was these interactions and their consequences that led to the loss of leak tightness of the containment at Chernobyl and Fukushima.

The main safety rule, in force today, is that, in the event of a severe accident, there are no consequences for the surrounding populations, that is, no radioactive releases into the environment. Thus, the design of conventional reactors must meet stringent requirements.

For an MSR, it is difficult to define a severe accident. Studies have shown that the most severe accidents imaginable seem to be either an uncontrolled rise in temperature with progressive loss of the structures containing the salt/fuel, or quite simply a leak from this same structure. In both cases, the nuclear reaction stops because the reactor is no longer in the reactive configuration. The second barrier provided for this purpose contains the salt, which will freeze, trapping most of the products, in particular the cesium. The third barrier, that is, the containment, ensures the tightness of the unit, as there is no release of energy, particularly since the combustible salt is not under pressure.

As there is no pressure in the salt circuits, it is possible to avoid energy release in the event of leaks.

This advantage specific to MSRs is an important safety element and improves the acceptance of these reactors by the population.

The advantages of MSRs in terms of the fuel cycle

A fast MSR can operate with a wide variety of fuels: natural or depleted or even enriched uranium, thorium, plutonium, actinides, etc. The only constraint is that this mixture of fissile, fertile, or even waste must reach criticality to guarantee operation. This flexibility is a huge asset since it allows all types of isotopes to be mixed in the fuel salt.

In the case of the U/Pu cycle, Pu is the preferred fissile element because it is already available in large quantities; similarly, depleted or reprocessed uranium serves as a fertile element. This U/Pu cycle has the following advantages:

- This is the best-known regenerative cycle, from the experience of sodium-cooled fast reactors (SFRs).
- France already has skills and operational industrial equipment.
- France has more than 300,000 tons of depleted uranium stored and available.
- If thorium (fertile element) is introduced, the production of ^{233}U gradually provides an additional fissile element. Note, however, that the solubility of Pu in a fuel salt containing thorium is lower.

The U/Pu cycle is therefore the one that seems to be a priority for MSRs in the French context.

Note, however, that the thorium cycle has several advantages:

- Thorium is also available (about 8500 tons in France and much more abroad).
- The Th/U cycle is one of only two regenerative cycles available.
- Thorium produces ^{233}U, which is an excellent fissile.
- This is a cycle that could take over after the depletion of plutonium stocks.

This cycle works with 20% enriched uranium, which makes the export and circulation of this type of reactor easier in terms of nonproliferation.

Therefore, it is an interesting candidate for further study.

The fertile elements currently available in France are uranium (natural, enriched, depleted, or reprocessed) and thorium. But if a separation of minor actinides becomes possible during reprocessing, these elements also become candidates for incineration in an MSR. During the salt reprocessing phase in the reactor unit, it suffices to put them back into the core (with the Pu and the U).

It should be noted that it is possible to start a reactor with various mixtures of fissile depending on availability, which makes the concept very flexible depending on the conditions at the time. Likewise, the power supply during operation has the same flexibility, the main constraint being maintaining criticality.

For example: in the EVOL deliverable (European project), the starting composition includes 77.5% LiF, 6.6% ThF_4, 12.3% UF_4 including 13% enriched uranium and 3.6% transuranics ($TRUF_3$: uranium oxide irradiated at 60 GWd/t, cooled for 5 years) (Fig. 9).

Finally, let us specify that, if we do not require regeneration, it is enough to not add any fertile element; in this case, we have a "burner" type reactor whose only function is the consumption of fissile isotopes in the fast spectrum.

Incineration

MSRs have interesting characteristics in terms of incinerating the actinides produced by the PWR sector, and calculations have been made for the use of different mixtures containing enriched uranium, thorium, ^{233}U, and minor actinides to start the reactor (see the European EVOL project).

The big advantage over solid fuel reactors is that they avoid long (and expensive) manufacturing/storage/use/decay/reprocessing/transport/etc. cycles of fuel (at best, 7 years outside the core for 5 years in the fast neutron reactor (FNR) core).

Moreover, in SFRs, it is very difficult to manufacture and handle a solid fuel containing minor actinides. In addition, products such as americium are very annoying neutron poisons in the reactor core, which require the manufacture of special assemblies for their incineration.

Conversely, in an MSR, the actinides are initially dissolved in the salt and their quantity decreases during the operation of the reactor. Salt reprocessing aims to leave them in the combustible salt to be incinerated. Better still, successive additions are possible to gradually increase the quantities incinerated.

For these reasons, the incineration of minor actinides seems much simpler in MSRs than in SFRs.

In conclusion, the fuel cycle has great flexibility in an MSR compared with in an FNR. In an FNR, it is necessary to manufacture assemblies with mixed oxide (MOX), introduce them for several cycles in the reactor, let them cool for a few cycles before extraction, cut them up for reprocessing (creation of waste), and separate and recondition the products. Then, with these products a manufacturing cycle is resumed that has become complex with the high activity of these products. Incineration of minor actinides would add further complexity.

In an MSR, the products are solubilized in the salt at the start with great flexibility in the possible composition. During the reprocessing of the salts, the nonincinerated products are returned to the reactor. The cycle is much simpler and shorter, and generates less solid waste.

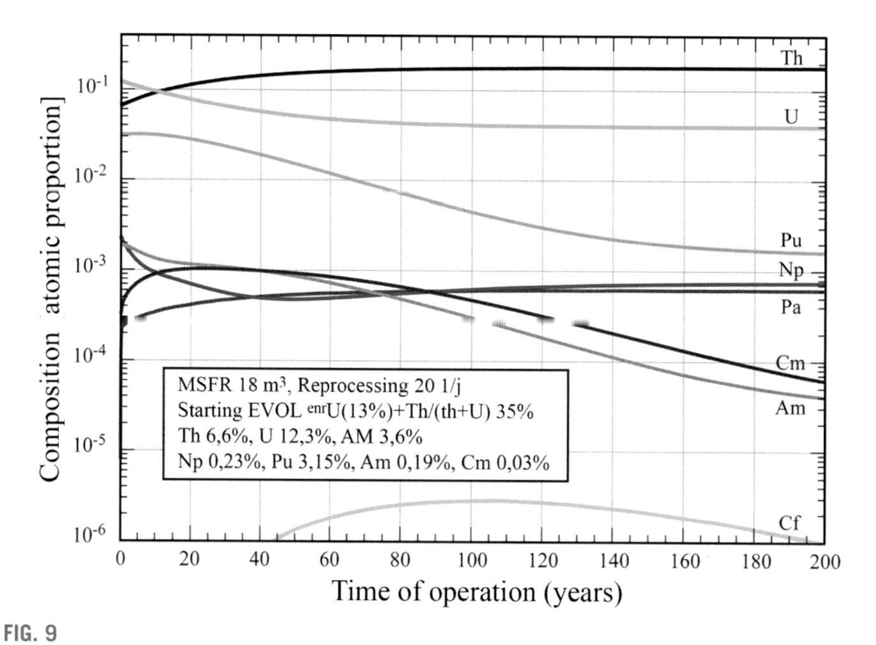

FIG. 9

Example of calculation of fuel evolution from an initial loading of the EVOL project.

The advantages of MSR in terms of network monitoring

The design of the MSR does not require control rods, for piloting or stopping.

Power is controlled by adjusting the temperature of the salt at the exchanger outlet, that is, at the core inlet. If we increase the power extraction at the level of the intermediate circuit, the core inlet temperature drops, which leads to a greater density of the fuel salt and, therefore, a reduction in neutron leakage. This causes an almost instantaneous increase in power in the core until a new equilibrium is reached between the temperature of the fuel salt and the neutron leakage.

If, on the other hand, the cooling is reduced, the salt at the exchanger outlet is hotter, and therefore less dense (more dilated). The neutron leaks are then greater, the reactivity drops, and the core "chokes" and causes a decrease in the power produced.

It should be noted that this counter-reaction mechanism by the expansion of the salt is very fast: it guarantees the response of the reactor to any variation in the inlet temperature of the salt in the central zone of the reactor. It is this characteristic, validated experimentally, that led to the use of MSRs in the design of nuclear planes in the 1950s, because piloting planes requires a high reaction rate.

In terms of reactor control, the excellent network monitoring possible with MSRs is a great advantage. Simulations show for the MSFR a passage from 2 to 3 GWt in the order of a minute (not achievable with a solid fuel). This very fast thermal counter-reaction makes several points possible:

- The power produced perfectly follows the power extracted.
- Load monitoring is controlled by the power drawn by the electrical network, and therefore by the speed of the pumps in the intermediate salt and combustible salt circuits.
- From a thermomechanical point of view, the reactor is suitable for this load monitoring with almost constant wall temperatures.
- Since the minimum/maximum temperature difference in the core is low even in the event of a variation in the power extracted by the exchangers, the thermal stresses on the structures are low, ensuring the longevity of the structures.

We therefore have a very "flexible" reactor that is well suited to load monitoring.

In addition, calculations have been carried out for a theoretical accident: if the reactor is at an initial power of 0.1 GWt, calculations show that, in the event of almost instantaneous overcooling (i.e., in 1 s), we observe a power increase to 3 GWt; the powers produced and extracted from the MSFR nevertheless remain balanced with a slight increase in the temperature of the fuel salt. Even in the case of very fast kinetics, the physics of the liquid fuel repositions the reactor in a new state of equilibrium (as for a normal piloting phase).

Obviously, this case does not detail how such cooling is physically possible. To optimize safety, it will be necessary to analyze all the situations where untimely cooling can occur (see safety analyses later in this chapter) and to look at the consequences on the operation of the reactor. Nevertheless, these studies show that load following (in particular due to the increase in the share of renewable energies) does not pose a problem in terms of the production of thermal power and the integrity of the structures linked to the fuel.

Finally, let us note several specific points of reactor control:

Startup of the reactor

- The startup requires a subcritical approach by filling the core and measuring the amplification of the neutron source. At mid-filling height, the simulations in the case of the MSFR show that we should be able to estimate the final reactivity at ~250 pcm.
- The filling continues with adjustment, if necessary, of the injected compositions.
- The divergence can take place either with circulation of the salt and corresponding adjustment of the inlet temperature, or with a filling made subcritical by an absorbent device (e.g., an absorbent reflective element) that is removed for the divergence.

Reactor shutdown

- The shutdown of the cooling leads to a shutdown of the reactor, where the power emitted is almost zero, apart from the residual power.
- A "safe" shutdown for a solid fuel reactor is a situation where the control rods have fallen, making any core criticality impossible. For an MSR, safe shutdown corresponds to a situation where the core is critical at almost zero fission power (critical mock-up type core).
- Residual heat removal systems must be available to manage temperature changes.
- A rapid emptying system remains necessary, particularly for accidental cases involving leaks.

Finally, a study of operation on natural convection of the fuel (as retained in the Terrestrial Energy project) would be interesting, given the simplifications involved and the general layout of the MSFR, which lends itself well to it. However, the overall control becomes more complex, with the loss of the setting parameter, which was the fuel pump speed, and the loss of control of a corresponding imposed flow rate in the core. Operation in natural convection requires accepting a much lower specific power, which drastically increases the initial inventory per electrical GW. This avenue may prove to be interesting to study with a view to simplifying the systems and ensuring robustness, in particular for small reactors.

In conclusion, these reactors are driven by power extraction in the intermediate circuit. This allows very rapid transitions from nominal power to reduced power and should enable excellent network tracking.

The potential advantages of MSRs in terms of compactness

Finally, these reactors are compact in terms of volume-to-power ratio, which should be a source of potential savings.

Situation of MSRs in the world in 2021

At the global level, there is today, in 2021, a clear renewed interest in MSRs, with many projects in different countries. The situation is highly evolving. China, Russia, and India have implemented state programs, whereas in the Anglo-Saxon world there a large number of startups making proposals on the market. All these programs are aimed primarily at producing demonstration reactors to answer the main technical questions and validate the calculation codes used.

For startups, a good criterion of maturity is the level of certification by the safety authorities. In this area, Terrestrial Energy (in Canada) and Molten Chloride Reactor Experiment (MCRE), in discussion with the Nuclear Regulatory Commission, seem to be in the lead at the end of 2021.

Chinese program

China is the country with the largest and most structured MSR program.

All resources have been allocated to a research center, SINAP, located in Shanghai, which develops the entire concept in all its forms: research on the concept, codes, materials, technology, reprocessing, etc. Around 1000 people are estimated to work in this center, with operational molten salt loops, technological research, and the announcement of a small prototype.

As usual, China starts from the initial basis of the MSRE concept, and with the same initial options, namely a thermal reactor using fluoride salts with graphite moderator. Initially, thorium seems to be used, which seems logical given the availability of uranium and thorium in the country (Figs. 10 and 11).

Currently, two demonstration reactors are under construction in Gobi Desert (Figs. 12 and 13).

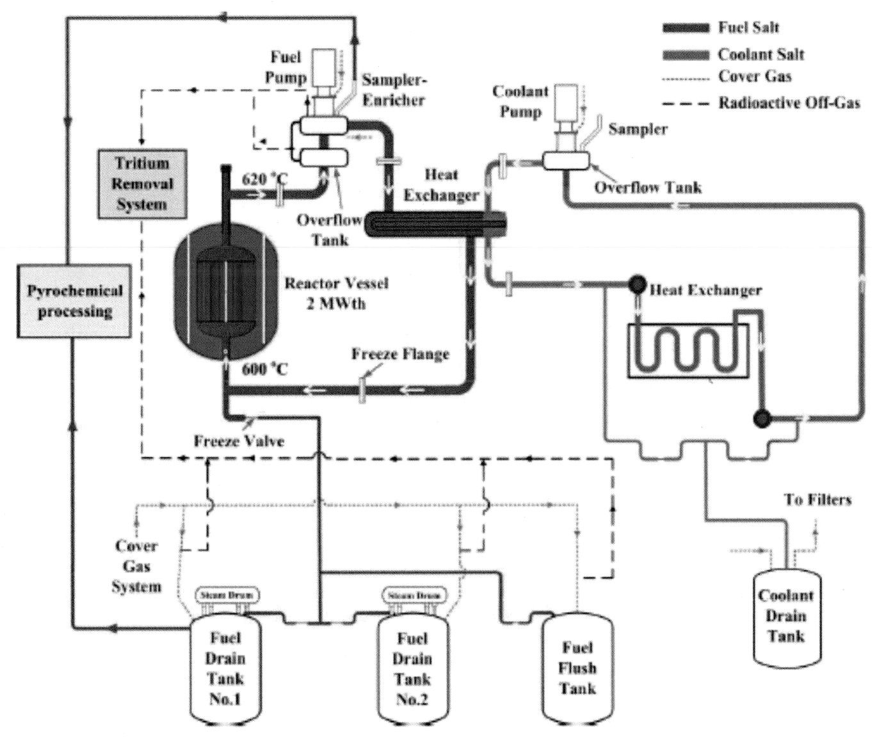

FIG. 10

View of the Chinese 2 MWt reactor project.

FIG. 11

View of the reactor block.

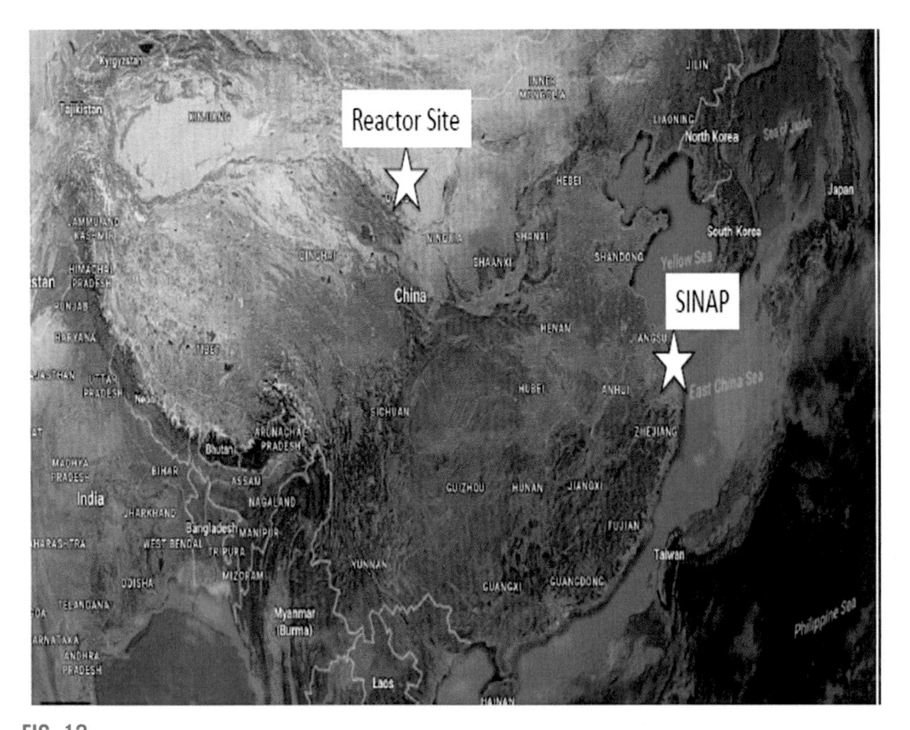

FIG. 12

View of the location of ongoing worksites.

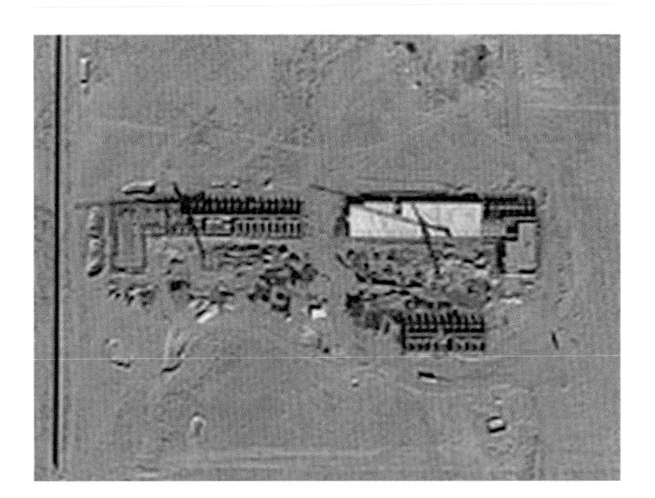

FIG. 13

Satellite view of the site at the start of 2020.

Russian program

In Russia, a small team is working at the Kurchatov Institute on the 1000 MWe reactor MOSART project. Here too, research is carried out around the concept in various fields. Research is continuing to obtain materials that withstand higher temperatures without corrosion. In the field of reprocessing, research is also ongoing. Valuable basic research has been carried out on the dissolution values of the various products in salts (Fig. 14).

The MOSART concept is more innovative: it is a fast reactor with fluoride salts, aiming to use available waste (minor actinides) while integrating the possibilities of thorium.

Some provisions are specific, such as the use, at core level, of a graphite wall to act as neutron reflectors and vessel protectors to increase its lifetime (Fig. 15).

On the basis of these studies, Rosatom announced in December 2019 the construction of an experimental molten salt reactor near Krasnoyarsk to incinerate minor actinides.

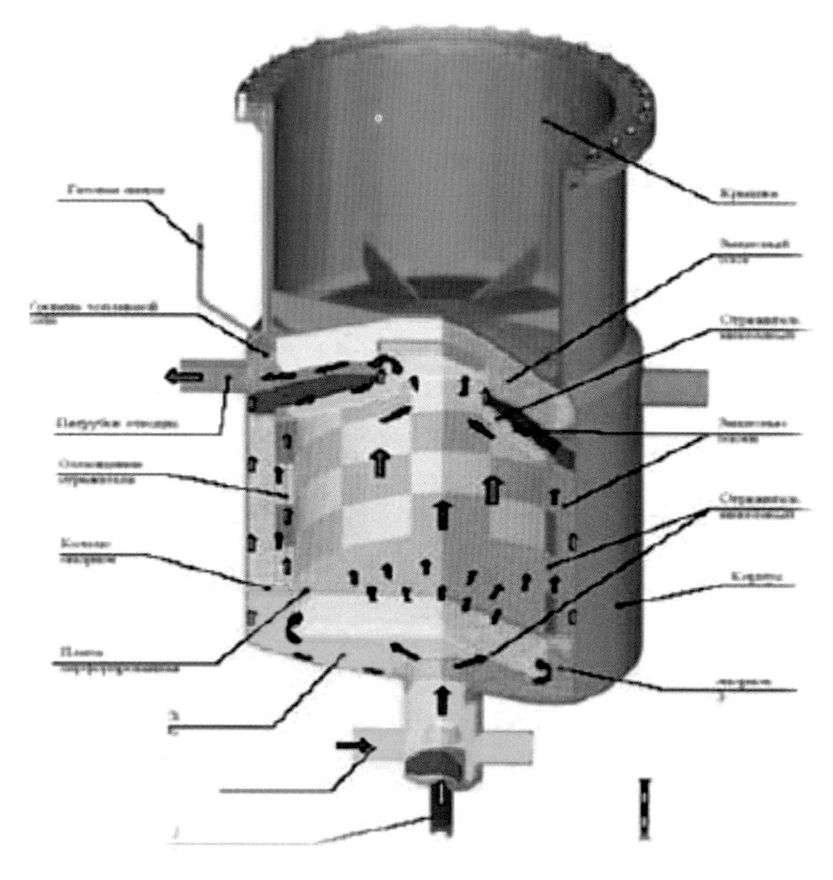

FIG. 14

View of the MOSART reactor block.

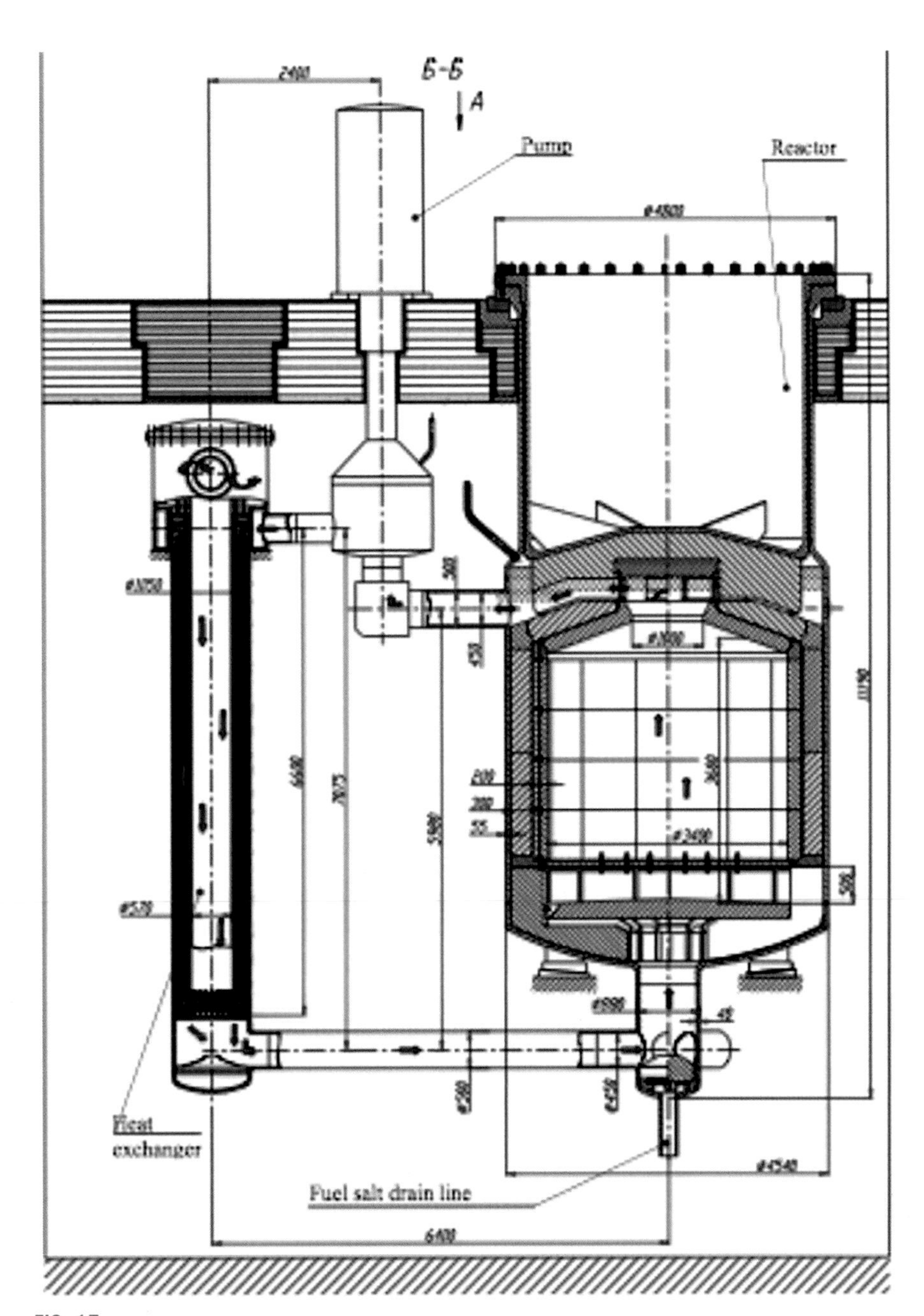

FIG. 15

Sectional view of the MOSART project.

Indian program

A more modest national program exists. It is a fast reactor, with fluoride salts, that recovers thorium, therefore consistent with the national strategy for closing the cycle and using thorium.

Attached is an illustration of their 850 MWe project (Fig. 16).

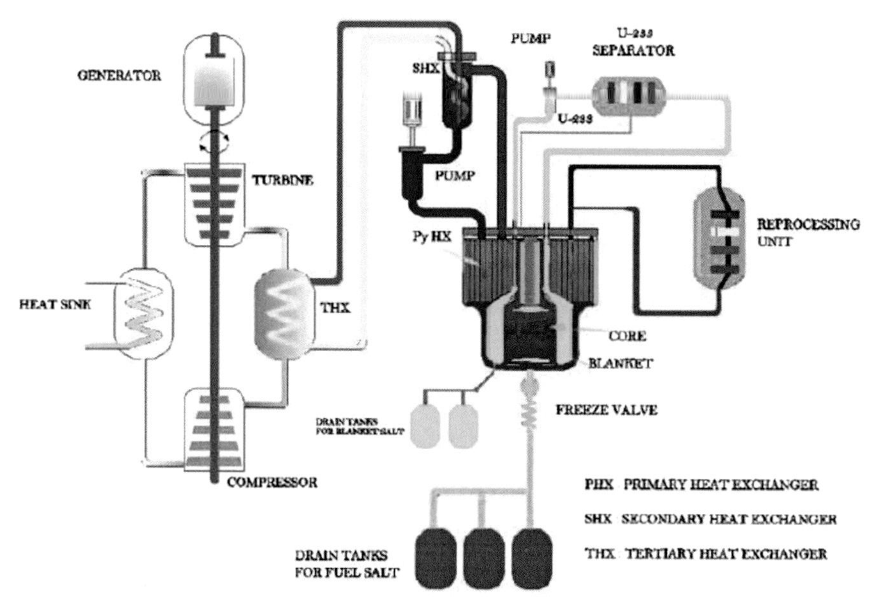

FIG. 16

View of the 850 MWe Indian project.

Anglo-Saxon startup projects

In the Anglo-Saxon world (the United States, Canada, the United Kingdom), many startups now offer reactor concepts using molten salts. Here we present of a list that is not necessarily exhaustive and is still evolving.

Note that some use molten salt only as a coolant and continue to use solid fuel, with clad. For example, MOLTEX combines a heat transfer salt with a solid assembly whose internal fuel becomes liquid in operation. Another example is Kairos Power, using salt as cooling fluid with HTR fuel TRISO.

Among the liquid fuel reactor projects are:

- Integral Molten Salt Reactor 400 (Terrestrial Energy, Canada) thermal.
- Molten Chloride Salt Fast Reactor (TerraPower, United States) fast.
- Thermal Transatomic Power Reactor (Transatomic + MIT, United States).
- ThorCon (Martingale, United States) thermal.
- Liquid Fluoride Thorium Reactor (Flibe Energy, United States) thermal.
- Molten Chloride Salt Fast Reactor (Elysium industries, United States) fast.

- Seaborg Waste Burner (Seaborg Technologies, Denmark) thermal.
- Copenhagen Atomics Waste Burner (Copenhagen Atomics, Denmark) thermal.
- Thorenco Process Heat Reactor.

These startups often have fairly limited resources, do little basic development, benefit from state aid (e.g., the Department of Energy (DOE)'s GAIN program), and also operate on the principle of filing patents or a license for which they are then marketable. However, the number of developing projects shows the dynamism of research and development on MSR, and illustrates the interests of the sector of molten fuel.

Of those developing a liquid fuel concept, the most advanced are discussed below.

Terrestrial energy's integral molten salt reactor in Canada

This project is a 400 MWt thermal reactor based on fluorides and using graphite. It is a concept with passive capacities for the evacuation of residual heat, burning uranium (low enriched uranium). The concept developed is passive (natural convection) and directly inspired by the MSRE of which it is intended as a direct extension.

It is the most advanced MSR in North America in terms of licensing, with the Canadian safety authorities.

It is a battery concept without reprocessing, where tank and salt are replaced every 7 years.

Its implementation is planned in Alberta for the use of heat for oilsands processing (Fig. 17).

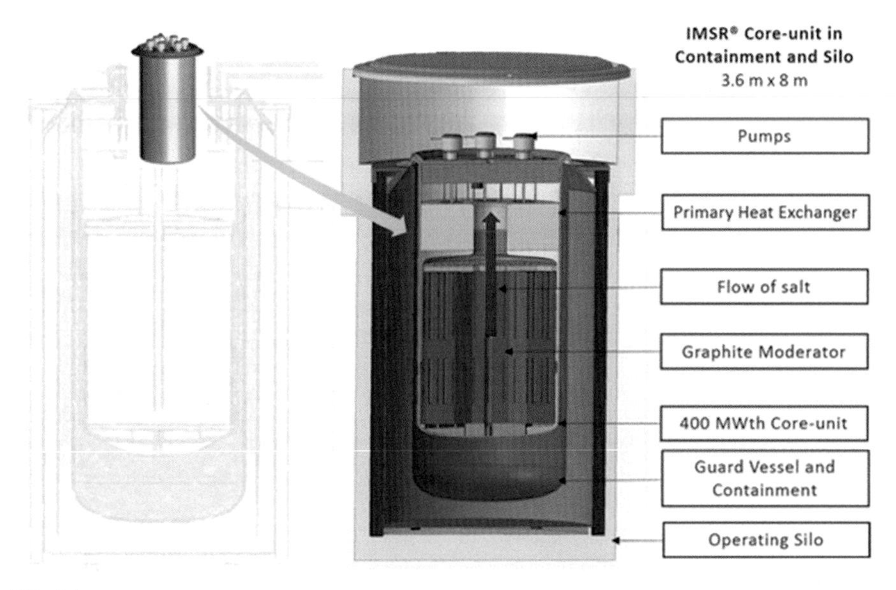

FIG. 17

View of terrestrial energy project.

The TerraPower molten chloride fast reactor

TerraPower presents a prototype reactor of 30 MWt, fast but with chloride salts. This is the project that comes closest to the concept studied in France (chlorides and U/Pu cycle). It is strongly supported by the DOE, for the short-term development of a demonstration reactor, the Molten Chloride Fast Reactor of TerraPower.

The reactor was studied with two regenerative cycles, U/Pu and Th/U, and is developed in cooperation with Southern Company Oak Ridge National Laboratory, Electric Power Research Institute, and Vanderbilt, who are providing technical support (Fig. 18).

The molten chloride salt fast reactor of elysium industry

As this 1000 MWe reactor uses a closed fuel cycle and spent fuel, a fast spectrum and a chloride salt were chosen (Fig. 19).

The transatomic power reactor (project of MIT)

This 1250 MWt project resembles an MSRE (thermal reactor and fluorides) but replaces the graphite with zirconium hydride (Fig. 20).

Thorcon power

Here we are dealing with a thermal project similar to MSRE, but in the form of modular concepts (twin reactor configuration). Fuel salt and moderator graphite are replaced every 4 years (Fig. 21).

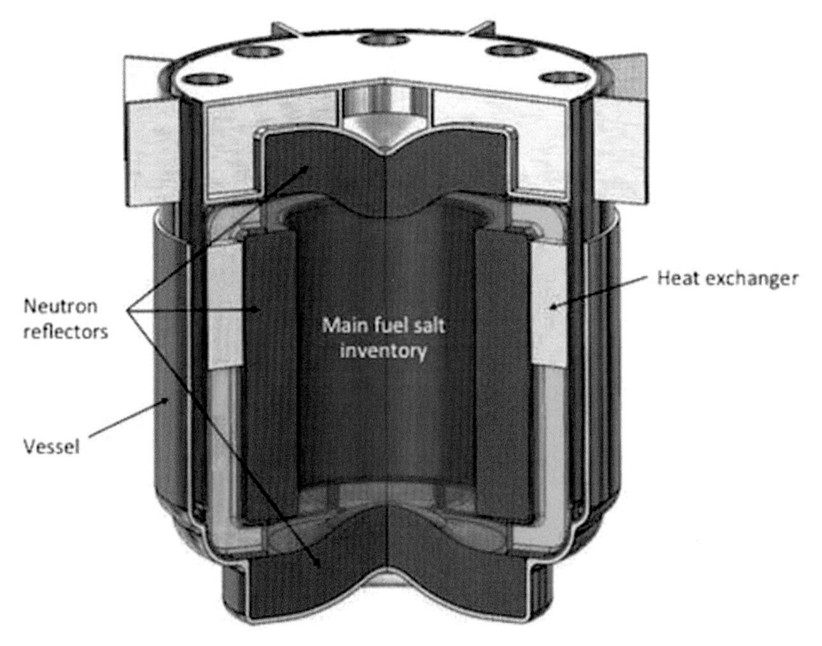

FIG. 18

View of TerraPower project.

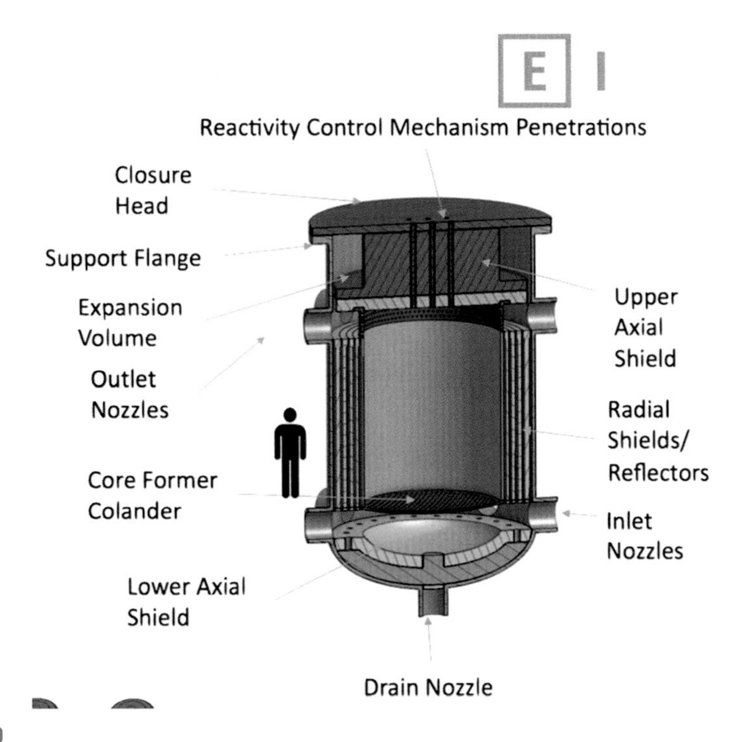

FIG. 19

The molten salt chloride fast reactor of elysium industry.

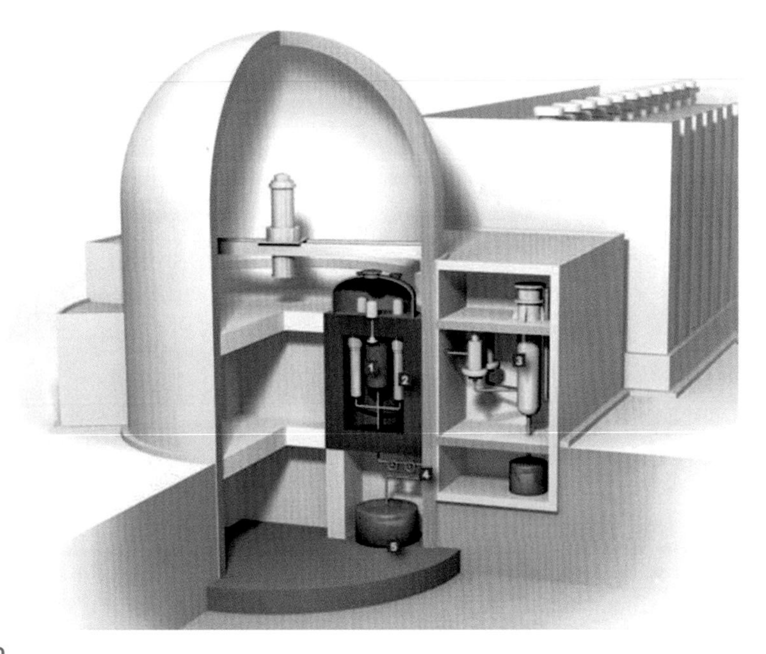

FIG. 20

View of transatomic project.

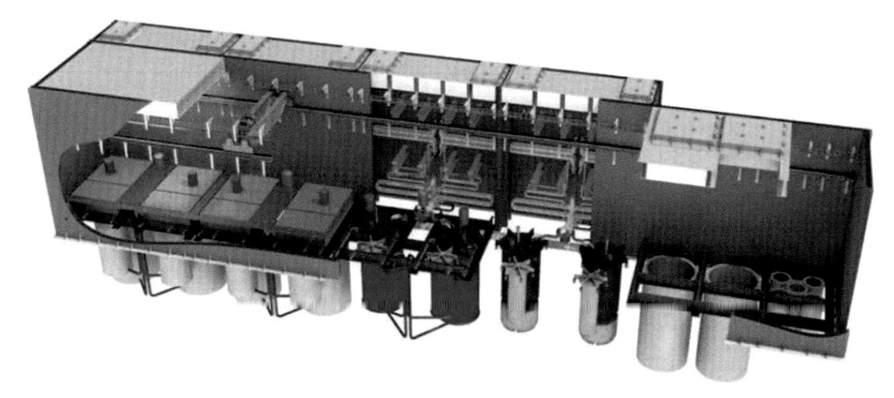

FIG. 21

View of Thorcon project.

Liquid fluoride thorium reactor of flibe energy

This project uses a thermal concept similar to the MSRE, but more focused on the use of thorium (thorium breeder fuel cycle/graphite moderated/thermal spectrum/ LiF-BeF2-UF4 fuel) (Fig. 22).

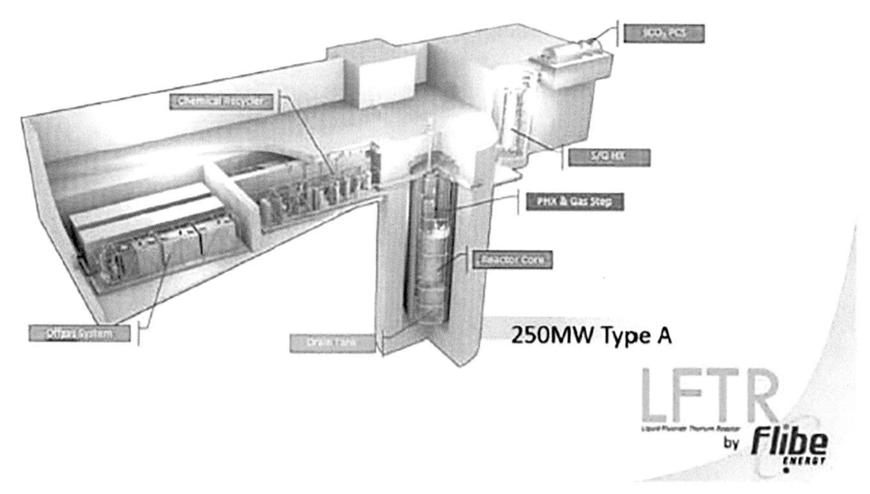

FIG. 22

View of liquid fluoride thorium reactor project.

Analysis of the different concepts

There is great variety in the technical choices proposed, highlighting the current lack of a consensus solution but also showing the versatility of the concept.

Many thermal power concepts are similar to MSRE, with a moderator (often graphite), and fast reactor concepts have all the potential advantages of fast reactors, particularly in terms of cycle closures.

Regarding the choice of salts, fluorides tend to be retained for thermal concepts based on MSRE feedback experience, and chlorides for U/Pu cycles.

The U/Pu cycle is commonly used for spent fuel, but the thorium cycle is another option, as well as the incineration of actinide-type waste.

The reprocessing of liquid fuel, when planned, is presented either on-line or in batch.

Depending on the desired application, several types of reactor are proposed.

The thorium cycle is the focus of many studies. As thorium is not fissile, the cycle begins with enriched uranium as fissile and thorium as fertile. This operation produces uranium-233, which is an excellent fissile. In the longer term, this uranium-233 can be extracted, making it possible to establish a self-supporting cycle with thorium. In this case, the salts used are generally fluorides. This is based on the MSRE experience of carrying out the second phase of operation after extraction of ^{233}U; however, this has never been done using thorium.

The fissile materials that can be used today are essentially either enriched uranium or plutonium when it is available. Current spent fuel reprocessing methods could lead to the subsequent disposal of other fissile elements, such as certain minor actinides of the americium type. The use of plutonium is, of course, preferred in the scheme of a fast reactor cycle. The operation of a reactor initially loaded with plutonium and depleted uranium is naturally almost isogenerating. An example calculation with a fluoride salt has been made in Ref. [4].

Unfortunately, the solubility of Pu is low in these salts, leading in cold transients to risks of plutonium deposition with the corresponding possibilities of local criticality (see Ref. [5]). For this reason, all MSR projects based on a cycle with plutonium use chlorides, where the solubility of Pu is sufficient in all temperature ranges. For neutronic reasons (see Ref. [5]), NaCl is generally retained. The other possible salts are less transparent to neutrons and have less favorable neutron balances.

It should be noted, however, that the ^{37}Cl isotope is a neutron poison [6] and that an isotopic separation is necessary on the initial salt of the first loading, to only load ^{35}Cl.

Special situation of France

The case of France is particular because the country has around 340,000 tons of depleted uranium through its enrichment activities, a stock that is increasing by around 7000 tons each year. France also has, through its reprocessing activities, a few hundred tons of plutonium stocks, which increase each year by about 10 tons. If not used, these products become additional nuclear waste.

MSR studies therefore focus primarily on depleted uranium/plutonium cycles, making it possible to use these products. These produces are waste stored in government places and speculation on their cost is today no possible. To give an order of magnitude, the first calculations carried out [4] show for a 1400 MWe MSR reactor that the initial loading would require approximately 9.2 tons of plutonium and 35 tons of depleted uranium. Three calculation cases are presented using depleted uranium, and successively Pu from reprocessed PWR fuel, then Pu from reprocessed spent MOX, and finally, Pu and minor actinides from reprocessed spent MOX with separation of minor actinides. The results show great neutron flexibility with, in all three cases, a consumption of natural uranium of around 1.15 tons/year and of Pu of around 30 kg/year. We are therefore almost in isogenerator mode. If one had wanted to study a Pu burner option, it would have been necessary to use Pu without uranium. Note, however, that these calculations were made with fluoride salts and that the use of chloride salts could modify these orders of magnitude.

These calculations show that in operation the reactor produces almost as much plutonium as it consumes. The addition of blankets would be necessary to make it breeder. It suffices to periodically add depleted uranium to maintain operation. Annual consumption is approximately 1.2 tons, that is, for a 100 GWe fleet less than 90 tons/year. This allows more than 4000 years of operation with the stock of depleted uranium available in France!

In 2050, the quantity of Pu available would be more than 700 tons. It is estimated that there is enough to operate an MSR isogenerator park of around 100 GWe. Adding blankets would make it possible to develop a fast breeder and further increase the quantities of plutonium available.

It should also be noted that stocks of plutonium exist in a few other countries. The stocks declared in 2013 were the following:

- Russia: 178 tons
- United Kingdom: 107.2 tons
- United States: 88.3 tons
- Japan: 47.1 tons
- Rest of the world: 13 tons

This explains why other countries are also interested in this U/Pu cycle, which in its fast version has various advantages: no need for uranium mines or enrichment plants, operation with available waste, and operation assured for thousands of years, ultimately producing easily manageable short-lived waste.

It should be remembered that, mainly for reasons of Pu solubility in fluoride salts, U/Pu cycle reactors generally use chloride salts. This choice has technical consequences that we will discuss later.

However, a more global reflection, that is, at the level of the planet, shows that these quantities of plutonium are not sufficient to ensure the energy demand projections toward the end of the century with these U/Pu cycle MSRs. The thorium cycle could then work in parallel. This cycle requires relatively available and abundant thorium as fertile, and, as fissile, 20% enriched uranium. We see that in the first stage we are not in the virtuous cycle of the fast reactors since we need uranium mines and enrichment units. On the other hand, the operation creates an isotope of uranium, ^{233}U, that is an excellent fissile. The principle would be to extract this fissile element during salt reprocessing, to initiate isogenerator reactors based on a thorium/^{233}U cycle. These reactors generally use fluoride salts where thorium has good solubility.

Currently, there are no fuel reprocessing plants capable of carrying out this extraction in an industrial manner and with sufficient efficiency. It should be recalled, however, that this extraction has been carried out experimentally with spent thorium fuel for the second operating phase of the MSRE and that it is therefore theoretically possible, in particular by pyrochemical processes.

France is therefore continuing to invest in research efforts on this cycle, as it is a longer-term option that could take over when the plutonium reserves are exhausted, or in the process of being exhausted. In particular, the research also explores whether the addition of thorium to U/Pu cycle reactors, either as a blanket or in the salt, would make it possible to create preparatory stocks of ^{233}U.

For more than 15 years, the Centre national de la recherche scientifique (CNRS) (LPSC Grenoble) has carried out studies on a fast spectrum fluoride salt MSR project. The MSFR (Fig. 23) is designed to close the cycle by using the waste produced by the nuclear industry current (U and Pu) and to incinerate the actinides. Numerous calculations have been carried out with the thorium cycle, but this type of reactor is by design very flexible in terms of fuel and can therefore utilize the U/Pu cycle with some modifications, particular related to the less favorable thermal characteristics of chlorides.

This type of reactor is potentially capable of transforming all our waste (depleted U, reprocessed U, thorium, Pu, and even actinides) into energy, thus closing the cycle and minimizing the final waste. The chosen power is 3 GWt.

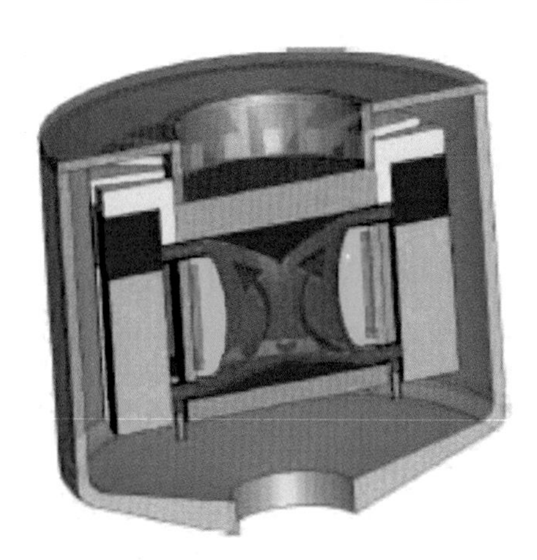

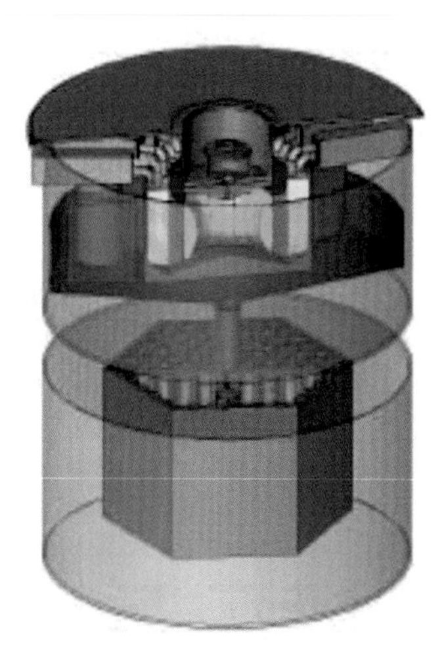

FIG. 23

View of the CNRS MSFR concept (LPSC Grenoble).

It is this MSFR reactor that served as a reference for the European project EVOL, then SAMOFAR. A new European project devoted to the safety analysis of the MSFR is currently underway (SAMOSAFER).

At the end of 2021, ORANO has stakes in the American MCRE project, Commissariat à l'energie atomique (CEA) and ORANO are working on a 300 MWe reactor project, and the French startup NAAREA is working on a 1–40 MW modular reactor.

The technical challenges of MSRs

Although not insurmountable, the technical challenges are numerous because feedback is limited to the MSRE, that is, a thermal reactor with fluoride salts and without reprocessing experience.

Choice of salt

The main constraints are that the salts should not activate and should have good resistance under irradiation, be transparent to neutrons, have good chemical stability at high temperature (> 1300°C), not produce radioelements that are difficult to manage, have low vapor pressure, have good thermal characteristics (conductivity and capacity), have good thermohydraulic properties, be able to solubilize fissile and fertile materials (uranium, plutonium, thorium), and facilitate reprocessing and redox control (see more detailed analysis in Appendix 3).

Two large families of salts can be used: chlorides and fluorides, which have their respective advantages and disadvantages.

Chlorides have the following main advantages:

- A melting temperature for some eutectics (around 400/500°C) lower than that of fluorides.
- Better solubility of Pu and fission products depending on the salt retained (e.g., binary $NaCl$-UCl_3).
- Easy solubility in water, which makes it a better candidate for the French PUREX reprocessing process. This also allows, in terms of maintenance, easier washing of the extracted components.

Nevertheless, chlorides have disadvantages that limit the above advantages:

- A harder spectrum, which can present certain neutron advantages for a fast version, but which leads to strong irradiation of materials close to the active zone (axial reflectors, fertile covers).
- A much more pronounced hygroscopic character (water absorption) than for fluorides and aggressive pitting corrosion.
- A larger "migration area," which leads to neutron leakage requiring either a larger core or the placement of reflectors.
- Worse thermal characteristics, which leads to larger exchangers.
- The need for extensive enrichment in ^{37}Cl.

Many projects propose fluorides for which we already have first feedback experience on the MSRE. These have good behavior under irradiation, good neutronic properties, and suitable melting temperatures if the eutectic is well chosen. They also simplify certain phases of reprocessing thanks to the volatility of UF_6.

However, they have some drawbacks:

- The operating temperatures are higher (between 600°C and 700°C).
- Li-6, a neutron poison, generates tritium. A thorough enrichment in Li-7 is necessary.
- The solubility of Pu is insufficient, which leads to a risk of deposition in certain operating ranges.
- There is a relative incompatibility with the PUREX reprocessing process available in France.

There is a wide range of possible fluorides: a mixture LiF and BeF_2 (FLiBe) was used on the MSRE, but KF and NaF are also possible.

Within the framework of the MSFR, and after selecting the fluorides, the following has been proposed: LiF-ThF_4-UF_4-$(TRU)F_3$, that is, a salt based on LiF (with about 22%mol of heavy nuclei). In this context, the following products have been avoided as a secondary component of fuel salt:

- BeF_2 because beryllium and its compounds are toxic (and rather to be avoided in Europe). It has the property of lowering the fusion temperature and increasing the thermalization of neutrons.
- KF because it is difficult to dehydrate before introduction into the reactor and can lead to the formation of gaseous K during high-temperature reductions.
- NaF because it does not bring much net advantage in terms of melting temperature (it lowers the temperature by 50°C) or solubility, and it is less chemically stable than LiF.
- ZrF4 lowers the melting temperature considerably and captures oxygen before the actinides, but it lowers the solubility of Pu, and its extraction is therefore necessary because its concentration increases (it is one of the main fission products).
- RbF is very transparent to neutrons and lowers the melting temperature, but may be problematic when reprocessing using any chemical reduction process.

Therefore, LiF is the final product retained by the CNRS as the salt base for the MSFR project. The melting point of the mixture with the heavy nuclei is 585°C. LiF has the advantage of being a monoconstituent base for the fuel salt thus formed, which simplifies the reprocessing of the fuel salt. On the other hand, a high enrichment in lithium-7 is necessary to minimize the production of tritium and reach criticality.

In conclusion, it is above all the choice of the fuel cycle that will determine the choice of salt. If we want to use plutonium in a U/Pu cycle, we will use chlorides, like the American MCRE.

If we have thorium available, like in China, we instead use fluorides, on the basis of the feedback experience of MSRE.

If we have a reprocessing device available that is better suited to one of the salts, we will tend to go in that direction.

A more detailed analysis of the choice of salt can be found in Appendix 5.

Neutronic

Initial studies of the MSFR focused on comparisons between thermal, epithermal, and fast spectra. Very quickly the choice fell on a fast spectrum, which better meets the needs of the French sector and has many advantages:

- Very negative feedback coefficients (advantages in terms of reactor safety and control).
- Better fission/transmutation capacity (benefits in terms of waste incineration).
- Much lower reprocessing needs thanks to the efficient incineration of certain neutron poisons such as americium (Fig. 24).

The MSFR is designed to contain a central zone of the tank where, owing to the critical geometry, a chain reaction with heat release takes place. The combustible fluid heats up and then continues its course to cool in the exchangers where the neutron flux is very low. At the outlet of the exchangers, the salt returns to the central zone (Fig. 24). The complete circuit is made up of independent sectors immersed in the tank: the salt never leaves the tank (Fig. 25–27).

A first particularity of the concept is the absence of control rods. Indeed, the very high coefficients of thermal counter-reaction induced by the expansion of the combustible salt and the absence of thermal inertia (unlike a solid fuel) make it possible to control the power very quickly. A control is then carried out by cooling the intermediate fluid.

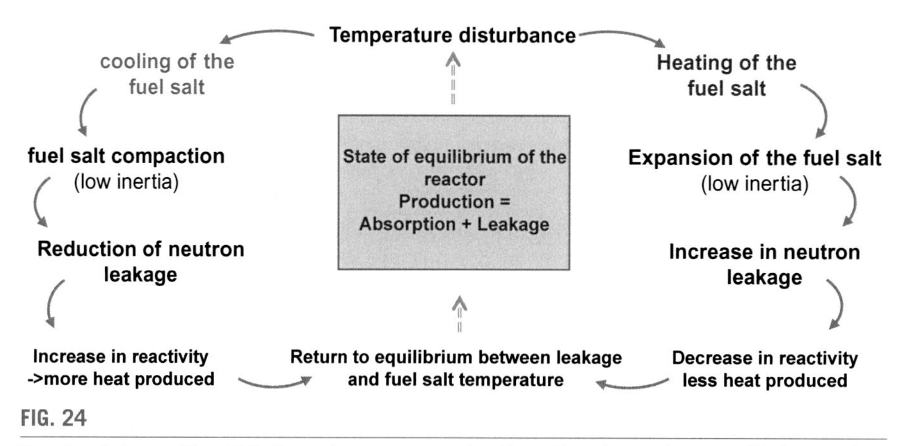

FIG. 24

Feedback loop of liquid fuel reactors guaranteeing the return to neutron equilibrium and allowing power control.

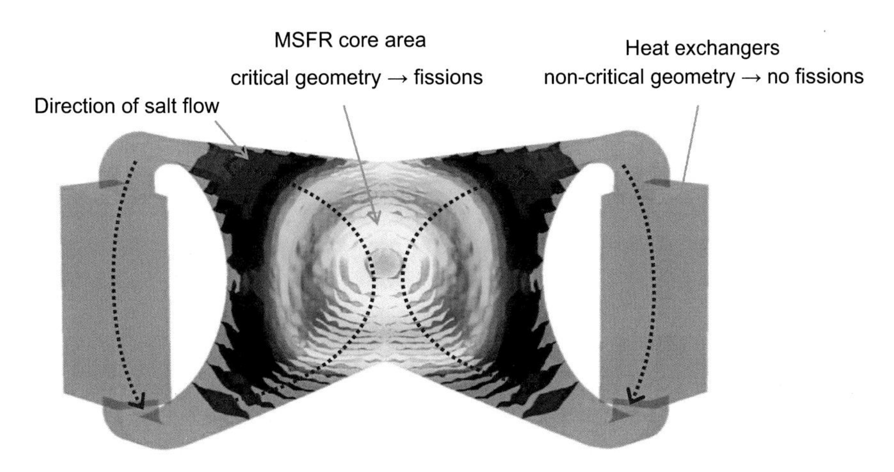

FIG. 25

Distribution of the fission rate in a sectional view of two sectors of the MSFR fuel circuit: the central zone allows the chain reaction owing to the critical geometry; in the exchangers, the geometry no longer allows the chain reaction.

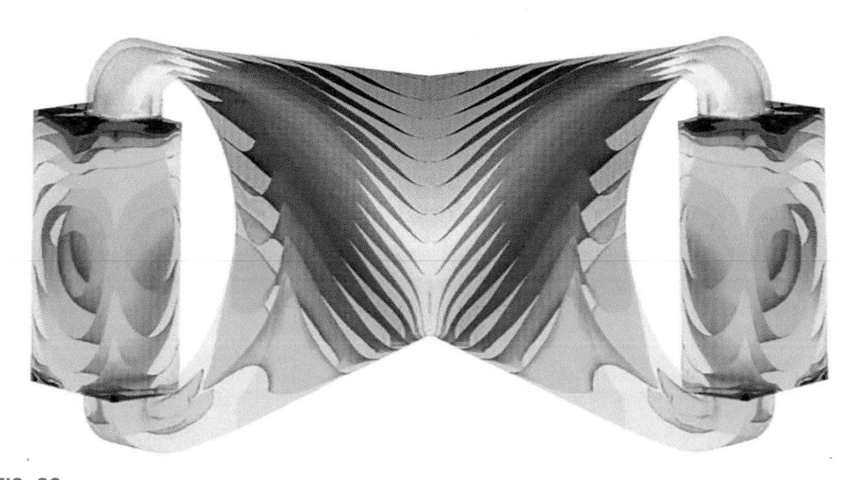

FIG. 26

Map of salt temperatures in a sectional view of two sectors of the MSFR fuel circuit: heating in the central zone, cooling in the exchangers.

A priori, there would be no need for control rods in terms of safety. Before placing one, it is necessary to determine its usefulness with regard to risk in the event of a malfunction. However, control rods can be useful to compensate for the loss of reactivity due to the consumption of fissile product.

Note that the shutdown situation of this reactor is similar to a model reactor at zero power: the salt inlet temperature corresponds to its outlet temperature. No power is drawn from the combustible salt, which remains hot.

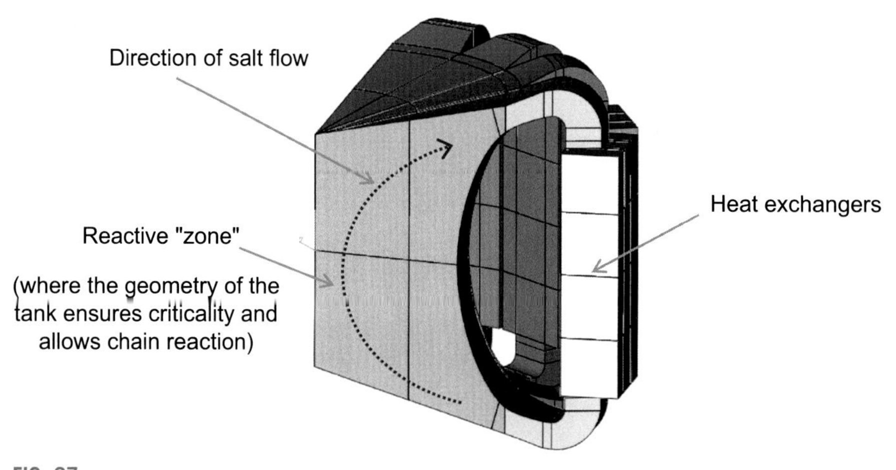

FIG. 27

View of several sectors (exchangers + pumps) immersed in the MSFR tank.

A relocation of the fuel salt, with subcritical geometry, by emptying the tank is also possible.

Calculations with neutronic/thermohydraulic coupling have been carried out to simulate the nominal state. Indeed, it is the strong expansion of the salt that induces the counter-reaction, allowing the piloting of the reactor. The calculations in progress have already taken this coupling into account, and it has also been validated experimentally on all liquid fuel reactors and experiments. This specificity of liquid fuel is one of the fundamental points of interest for the sector.

Finally, some of the delayed neutrons, necessary for the overall neutron stability, are entrained in the combustible fluid, which leaves the central zone to circulate in the exchangers. This point specific to MSRs must of course be considered in the calculations since it reduces the effective beta. It should be noted, however, that this reduction in the effective beta increases the speed of the counter-reactions and thus reinforces the stability of the reactor.

Several very interesting points emerge from the calculations already carried out:

- Efficiency in a reduced volume: we arrive for the MSFR (1400 MWe) at 18 m^3 with 330 W/cm^3 with an initial inventory of 3.5 t/GWe. It should be noted, however, that the Russian and American projects chose a lower power density. Indeed, the high power density makes the concept attractive in terms of compactness, use of the material, and therefore deployment capacity. But it introduces more constraints on the capacities of the exchangers to evacuate the power. The power density for MOSART is at least one-third of the value of the MSFR. The level of this power density is a design choice.
- For comparison, in a PWR, there are about 3 tons of ^{235}U for 1 GWe for a thermal spectrum. A fast spectrum molten salt reactor can mobilize no more material than a thermal spectrum reactor.

- High waste combustion efficiency.
- A major advantage of the fast spectrum compared with a thermal spectrum is the ability to operate without being poisoned by fission products (low reserve of reactivity required).
- The possibility of fissioning all the actinides (e.g., americium) since they are kept in the core. It should be noted that the low rate of actinide-delayed neutrons does not induce any consequences on the stability of the reactor (see remark on the low effective beta).
- The fast spectrum leads to much lower reprocessing needs (between 10 and 40 L/day). This important point will be discussed later in the chapter on reprocessing.

It should be noted that zirconium is a neutron poison and that its production can lead, by continuous accumulation, to difficulties after a few years. It will therefore be necessary to find a way to extract it during reprocessing.

It should also be noted that the design is based on the choice of the specific power (in W/cm^3), which determines the capacity of the exchangers to extract the heat produced. This remains the same regardless of the nominal power sought for the reactor.

The spectrum in the MSFR is a fast spectrum, a little more moderate than that of a sodium-cooled fast reactor owing to the use of the fluoride salt. The neutrons of several hundred keV are moved toward the upper epithermal zone. Using a chloride salt would lead to a harder spectrum (Fig. 28).

Finally, an interesting point for all MSRs is that a neutron excursion does not result in the melting of a solid core and its interaction with the cooling fluid, but in a simple expansion of the fluid of the central zone of the core, leading to an increase in neutron

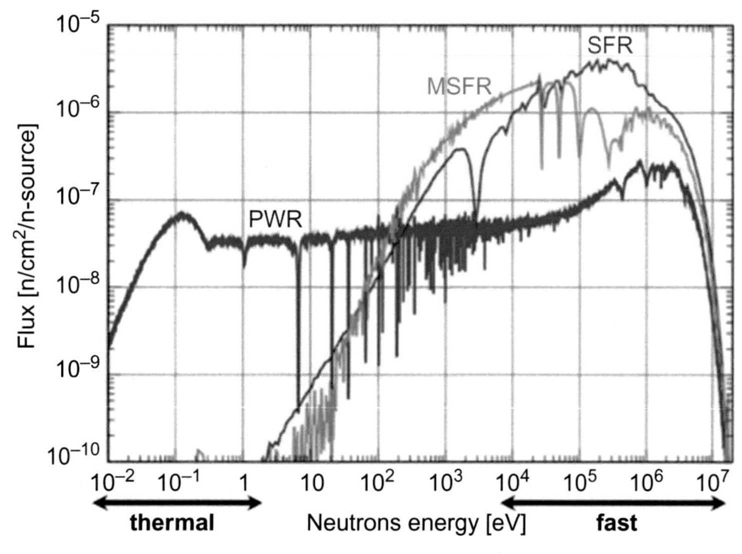

FIG. 28

MSFR neutron spectrum (green curve).

leakage and therefore a decrease in reactivity. Thus, subject to certain reservations (existence of a free surface outside the flow, etc.), the potential release of mechanical energy is almost nil, which is an advantage in terms of reactor safety and acceptability.

Note that this type of operation was tested and validated on the CEA's Silène reactor from 1974 to 2010. This reactor had a core that was an annular cylinder with a diameter of 36 cm. This cylinder was filled before an experiment with a solution of uranyl nitrate enriched to 93% in ^{235}U. A neutrophage control rod inserted in the central channel kept the reactor in critical mode despite the excess reactivity of the liquid fuel. Removing this bar allowed one to switch to a delayed or prompt critical mode, depending on the speed of extraction of the bar. This extraction speed allowed different behavior modes: level mode, free evolution mode, and salvo mode. In burst mode, which was the fastest extraction, a peak power of 1 GW was reached. During this phase, the rise in temperature and the creation of radiolysis bubbles led to significant feedback coefficients and rapid stabilization of the system.

Neutronics and isotope separation

It should be remembered that, in both cases, fluoride salts and chloride salts, isotopic separation is necessary.

Regarding fluoride salts, lithium-6 is a neutron poison and a tritium generator. It is therefore necessary to carry out a thorough isotopic separation (more than 99%) to keep only lithium-7. It should be noted that Électricité de France (EDF) already uses lithium enriched with lithium-7 to manage the pH of its waters.

Regarding chlorides, chlorine-35 generates chlorine-36, which is difficult to manage. It is also a troublesome neutron poison. China recommends an enrichment of 97% or more in chlorine-37 in their study of the MCFR [5].

Choice of material

Following the initial setbacks with Inconel on the ARE, the MSRE project developed a new material based on nickel and molybdenum: Hastelloy-N, which was used for the tank and the exchangers and for which we therefore have operating experience [1,2]. This feedback experience showed that, subject to control of the U^{4+}/U^{3+} redox potential, corrosion was limited, at 650°C, to approximately 2.5 μm/year.

In fact, the pure salt was not the corrosive agent but rather the pollutants (water, residual oxides, and fission products) dissolved in the salt.

The MSRE experience thus identified corrosion by tellurium [2]. Metallic Te forms with the Cr of the alloy an intermetallic compound that concentrates at the grain boundaries and weakens the alloy. To avoid this phenomenon, it suffices to reduce the metallic Te into telluride ions Te^{2-}, which is soluble in the salt. This reduction is ensured by the presence of a chemical buffer U^{4+}/U^{3+}. This buffer makes it possible to maintain the chemical potential of the salt in a range where corrosion is very limited. Moreover, fission reactions have an oxidative character. Indeed, when fissions cause a U^{4+} ion to disappear and be replaced by a set of fission products whose average charge is close to 3+, the medium becomes more oxidizing, requiring control of the redox potential to avoid corrosion.

Subject to control of the redox potential, we hope to be able to limit these corrosion problems.

It should also be noted that Hastelloy-N had at the time been certified (pressure and nuclear vessel) for operation up to 1300°F, that is, 704°C.

However, in addition to corrosion, a material with a fast spectrum is subjected to high irradiation levels and helium production (n, α reaction on the Ni), which weakens it. Although these effects are not well known, it is possible that high temperatures could allow amelioration of defects by diffusion and elimination of helium.

Other degradation factors are creep and cyclic fatigue linked to temperature fluctuations (stopping and restarting phases) and free level fluctuations.

For these reasons, the use of materials that would be even more efficient is also studied: new alloys based on nickel (EM819 from Aubert & Duval), with molybdenum and/or tungsten. The alloy with tungsten has been studied. Aubert & Duval produced the EM721 and EM722, and Russia tested the EM721 [6]. However, none of these new materials is available and validated.

Concerning the protective materials of the tank, graphite poses problems of volume change depending on temperature and fluency. Its use is therefore simpler in thermal flux than in fast flux. It was, however, retained by the Russians on the rapid MOSART project to protect the vessel material. Its lifespan is also limited (Terrestrial Energy in its thermal version has a duration of 7 years). It should also be noted that the management of irradiated graphite is still problematic (Fig. 29).

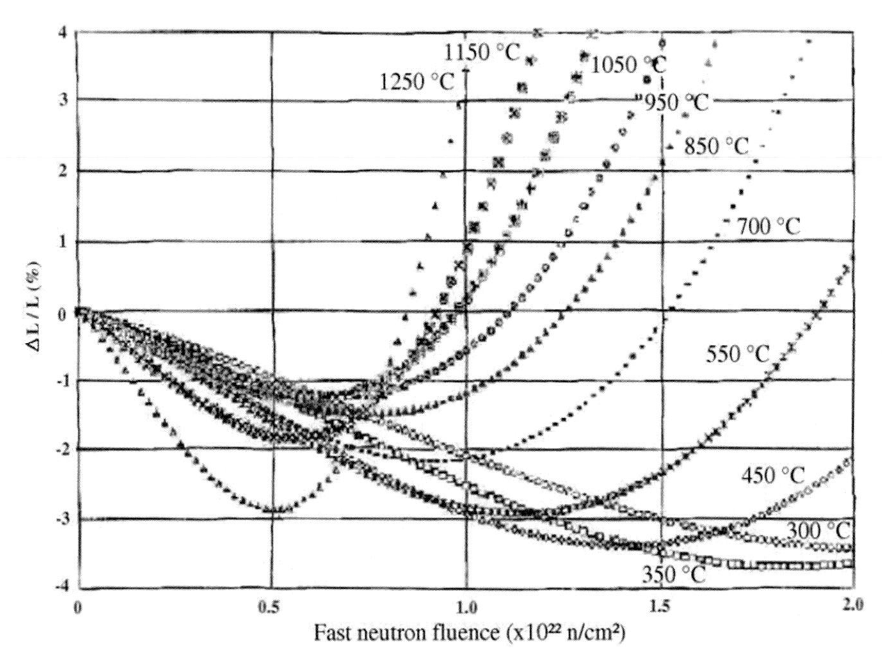

FIG. 29

Dimensional variation of graphite as a function of fluency and temperature.

Finally, SiC is another interesting material (ceramic type), for the thermal protection of the walls directly in contact with the fuel salt (axial reflectors and fertile cover) and to build the exchangers because it has good thermal conductivity. It has good mechanical strength up to 2000°C, good corrosion resistance if it is pure enough, and good resistance to irradiation. It should be recalled that it is used in HTRs as a coating for combustible pebbles, where it undergoes very strong irradiation without damage.

On the other hand, it poses certain industrial problems (mode of assembly, connection with metals, etc.) and at these temperatures risks rupture of the "brittle" type (propagation of the crack in the event of rupture, even if the material is very strong). It should be noted that the use of SiC fiber composites makes it possible to eliminate the risk of "brittle" type fracture. Finally, its thermal conductivity is much higher than that of steels (by about a factor of 5), which makes it an excellent candidate, for example, for heat exchangers. Modern additive manufacturing processes would also solve the problems of assembling a material that is not weldable.

If we stay in the field of additive manufacturing, we can note that some materials very recently developed in this context, such as graphene, have interesting characteristics.

If we compare available materials and salts, we arrive at the following table, which is not exhaustive (Fig. 30):

Material	Estimated working temperature	Main benefits	Major disadvantages	Doubts to clear
Hastelloy N	700°	Production and assembly of complex parts. Proven resistance to corrosion	Excessive creep at T° > 750°C	Resistance to irradiation +He at 700°C
Ni-Cr (Inconel 800)	1000°	Production and assembly of complex parts	Requires a reducing salt U_{3+} and Pu^{3+}	Behaviour under irradiation
Mo	>1700°C	Chemical and temperature resistance	Assemblies Fragile	
Mo-Re	1600°C	Chemical and temperature resistance	Transmutation of Re	Shaping Welding
SiC	>1700°C	Chemical and temperature resistance. Resistance to irradiation	Brittle break	Holding of assemblies

<700°C

>700°C

Cycle	Fuel	Melting temperature of the salt	Operating temperature of the materials
Th/U (MSFR)	LiF-ThF4-UF4	580	680/730°
Th/U (MSBR)	LiF-BeF2-ThF4-UF4	500	600/650°
U/Pu	NaCl-MgCl2-UCl-PuCl3	450	550/600°
U/Pu incinérateur	NaF-KF-UF4-PuF3	450	550/600°
U/Pu	LiF-UF3-PuF3	700	800/850°
U/Pu	LiF-UF4-PuF4	500	600/650°

FIG. 30

Comparative assessment of materials and salts.

In conclusion, today the only material officially available and validated seems to be the Hastelloy-N used on the MSRE (and its Chinese counterpart, the GH3535). But research is underway on promising materials, such as SiC.

Salt/fuel chemistry

Initially, the fissile products are dissolved in the retained salt, which allows the startup of the reactor with a well-known initial composition. A large number of products are then created by nuclear reactions during operation, creating an extremely complex overall composition.

Tables giving the solubility of the main products concerned exist, such as those published by the Kurchatov Institute and the JRC Karlsruhe. However, it is necessary to be able to monitor the real chemical composition of the combustible fluid over time.

In addition, the rise in concentration of certain fission products, such as lanthanides or zirconium, causes them to reach their solubility limit and leads to risks of deposition (possibly in co-deposition with PuF_3). This is one of the points that requires the reprocessing of the fuel salt for their extraction. In terms of safety, it will certainly be necessary to have in-line sampling in the fuel circuit, with continuous circulation (e.g., on a cold trap, of the type for in-line purification of sodium-cooled fast reactors). If this system works for MSR, it should allow one to be sure of where the products will be deposited and also to be able to extract and treat them (either for recycling or as waste). It should be noted that the operating temperature retained for the cold trap would then be the temperature to be taken into account for the initial solubilization and that this innovative concept would still need to be validated (e.g., on a molten salt loop).

Finally, the risks linked to the selective solidification of particular isotopes will have to be studied, as well as the procedures for reheating for melting and returning the salt to the fuel circuit during startup and operation.

In conclusion, the operation of a reactor will lead to the creation of many products in the fuel salt, which must be monitored and managed to avoid any untimely deposition.

Fission gas management

Neutron reactions lead to the continuous formation of fission gases and insoluble fission products. Some of these products are metallic and could lead to untimely deposits, particularly in the exchangers. Others are gaseous (mainly Kr, Xe) and will naturally escape to the free surface. It is best if they are extracted and processed continuously.

On the MSRE, sealing at the primary mechanical pump shaft was ensured by continuous injection of helium. It was this helium that carried the fission gases to their final treatment. Note in the REX MSRE some filter plugging and the strong influence of gas entrainment variations on core reactivity, due to the thermal spectrum since the gas contains absorbing fission products (nonexistent effect in fast spectrum). Note the enrichment of lithium to 99.995% in Li-7 to suppress the production of tritium (reaction $n+$Li-6).

On the MSRE, this extraction of gaseous fission products was done by bubbling with a neutral gas. Once this extraction is carried out, it is possible to carry out selective separations to be defined according to the gases produced. These separations are fairly standard provisions on all reactors (PWR, sodium-cooled fast reactor, etc.) in the event of solid fuel cladding ruptures. The separation processes must be defined, depending on the products to be separated, up to the final waste management method. This management can range from simple discharge (inactive gas with a low krypton type period), to storage in decay flasks for short-lived products, and to final storage for other products (activated carbons, etc.). Useful reference may be made to the existing separation systems at Phénix and Superphénix designed for the treatment of these gases, in the event of an accident situation with the simultaneous rupture of a number of clads.

Choice of intermediate fluid

An intermediate fluid is necessary to evacuate the heat from the intermediate exchangers to the energy conversion circuit (into water or gas).

This fluid must have the following characteristics:

- Liquid at operating temperatures with the necessary margins for solidification and vaporization.
- Withstand high levels of irradiation without activating. The activation data must also be confirmed (if there is radiolysis or creation of radioisotopes by activation, the management of this intermediate fluid is complexified).
- Not corrosive with respect to the materials used, in particular the material of the intermediate exchanger, which must support the combustible and intermediate fluids.
- Not cause major disturbances in the event of a leak onside the liquid fuel.
- Able to manage the consequences of a water leak in the intermediate circuit, in the case of a water/steam energy conversion system (detection and mitigation). The detection of entry of water vapor and management of the consequences remains, however, a delicate point, and requires further investigation.

For all these reasons, a liquid salt is generally used as an intermediate fluid. Studies carried out on MSBR have proposed sodium fluoroborate ($NaBF_4$ with 8% NaF), which has the advantage of not being activated and of capturing tritium.

Some, for fluoride salt reactors, have proposed using a mixture of fluorides (LiF-NaF-BeF_2, LiF-NaF-ZrF_4, LiF-NaF-BF_3, LiF-NaF-KF) [7,8].

The criterion relating to the risk of leakage into the fuel salt sometimes could lead to the choice of a salt identical to the salt retained as fuel salt.

It should also be noted that this assessment does not exclude the possibility of identifying a fluid other than a salt that also meets all the criteria.

The components

The reactor components must be able to operate in high temperature ranges (around 700°C) and without corrosion problems. Compared with fully integrated designs,

MSRs lend themselves well to a modular component view, which simplifies manufacturing. Experience on SFR components is also useful given the temperature ranges. Finally, the molten salt industry has interesting experience in the field.

Pumps

SFRs use mechanical pumps at high temperatures (400–500°C) and with certain technical solutions (such as hydrostatic bearing) that can be adapted to the MSR. These mechanical pumps were used on the MSRE. Adaptations, particularly at the level of materials, remain to be developed. Electromagnetic pumps are effective only with liquid metals. Diaphragm pumps are used in the molten salt industry but seem unsuitable. In conclusion, subject to benchmarking with these pumps, an adaptation of mechanical pumps today seems to be the best industrial solution. Note that the use of a magnetic drive (used in particular on the AP1000) would make it possible to avoid crossing the tank and thus maintain its tightness.

Exchangers

All types of industrial heat exchangers are suitable. The main problem is the choice of materials avoiding corrosion or deposit. Those allowing the extraction of a high power density are preferred, to minimize the volume of salt outside the critical zone. Thus, materials with high conductivity and compact plate exchangers are avenues to be studied.

Valves

Valves are necessary for the industrial operation of a reactor. For SFRs, there are valves with solidified seals or bellows valves. The principle of valves with solidified seals seems a priori not very adaptable to molten salts, because of the differences in mechanical properties of sodium and salt when they are frozen (solid sodium remains "flexible," whereas a salt becomes very hard). It would therefore be necessary either to develop bellows valves that can tolerate these temperatures and the fluid used, or to develop other valve concepts.

The loop tests, carried out in particular at the CNRS in Grenoble, mainly used pressure differences or passive devices to drain or transfer the molten salt. The Grenoble loop, however, used a specially designed mechanical valve that worked well.

Preheating

All filling and emptying procedures require preheating technologies for pipes and containers. Here, inspiration can be taken from SFR technologies, although the temperature levels to be reached are higher (melting temperature of around 100°C for sodium and 585°C for fluoride salt). Insulation and maintaining the temperature of the pipes are elements to be taken into account to avoid untimely freezing of the salt. This is a point that justifies continuing to consider low melting salts (ternary salt) for the intermediate circuit (e.g., the melting temperature of fluoroborate is 384°C).

Drain tanks

These tanks have several functions:

- Transfer of salt in normal operation for maintenance.
- Quick transfer in case of accident (e.g., circuit leak).
- Reception of salt patches at the beginning of the life of the reactor, and storage allowing filling for the startup of a reactor.

Here we find the functions of the SFR sodium storage tanks.

The transfer function is a reversible function, and the MSRE offers first feedback for their operation.

A single tank must be able (as on SFRs) to perform all these functions. Its design must make it possible to passively evacuate the residual heat but also to maintain (by preheating if necessary) the salt in the liquid state in order to be able to return it easily to its circuit.

Electricity generating circuit

If a classic "water-steam" type cycle is used for the production of electricity, the high temperatures will make it possible to achieve high yields (about 45% for the MSFR). Other solutions are possible, such as supercritical CO_2 circuits.

In conclusion: development work is necessary to identify suitable components. It should be noted that the washing/decontamination methods for the components extracted for maintenance have not yet been developed and are not necessarily simple. One solution is to treat these modular components directly as final waste (not reusable) and replace them with new components.

Instrumentation

A precise inventory of what exists in the industry using molten salts is needed; however, some instrumentation used in high-temperature SFRs can be used, subject to adaptation.

Temperature measurement

The principle of chromel/alumel thermocouples remains perfectly adapted. However, for thermocouples immersed in salt, the sheath must be designed with a suitable material.

Flowrate measurements

There is no obvious solution, because the electromagnetic flowmeters used on the SFRs are not suitable for salt.

Neutron measurements

The SFR neutron measurements outside the vessel are a priori usable in the same external context.

Level and pressure measurements

These remain to be carried out. However, robust electromechanical systems of the same type are possible.

Monitoring the chemical compositions of salt

It is essential to be able to follow the evolution of the chemical composition of the salt over time. Currently, given the dose levels, online monitoring by spectrometry seems difficult but not impossible and remains to be studied.

Discontinuous sampling should allow measurements according to a procedure that remains to be specified and validated. The method of sampling and route of these samples is to be specified. The samples could be installed at the level of the cold traps. The samples will be sent for reprocessing or returned directly to the tank.

Leak detection/dosimetry

The solidification of the salt and its nonreactivity with respect to air lead us to consider activity detectors that are already available and very sensitive to salt activation.

In the parts where the salt is not active, the leak detection modes require other types of detectors (e.g., by using the conductivity of the salt on the mode of the detectors used for the SFRs on the intermediate circuit).

In-service inspection

Here we are faced with a fairly unexplored field in terms of the needs and the means to meet them. This being said, the relative simplicity of the fuel circuit should allow a certain number of checks by external mode, such as wall thickness monitoring, by ultrasound, which should be sufficient. In this field, the nuclear industry has good experience.

Security approach and licensing

The strong specificities of the MSR must be taken into account when defining the safety approach, which must differ from the approach used for solid fuel reactors, taking into account the specificities of liquid fuel.

For example, in the specific context of MSR, it is necessary to redefine the notions of barrier, cases of severe accidents, or even safe stopping options. For the design of MSRs, It will be useful to highlight intrinsically favorable characteristics of the concept and try to take advantage of them in terms of simplicity.

On the other hand, the general approach to safety remains identical, namely the application of the principles of defense in depth. Thus, situations likely to lead to early or significant releases must be prevented, and in the event of occurrence, their consequences must remain limited and not lead to evacuation of surrounding populations.

The first definition of safety analysis was carried out in the latest European SAMOFAR project, starting from the specificities to be taken into account for liquid fuel reactors. A number of points requiring further study have been identified with regard to the three major aspects of safety: control of reactivity, containment, and evacuation of residual power.

Responsiveness control for an MSR:

- Intrinsic safety by dilation effect against reactivity accidents.
- Simplification of the architecture of the reactivity control (no command bar for the safety function)
- Total quantity in the core of fission products emitting delayed neutrons, low in absolute value and with an effect on the reactivity to be assessed in the study of transients (uncertainties included).
- Possible risks of reactivity insertion to identify and evaluate in the different states (including start/stop). For example: injection of excess fissile material, precipitation of fissile material, filling of fertile cavities with fissile material, cold shock, etc.
- Risk of criticality to be assessed throughout the installation (in particular, with regard to the risk of precipitation/crystallization).

Confinement

- The fuel circuit is not pressurized, and the salt presents no risk of a violent exothermic chemical reaction.
- Minimum number of barriers to be determined on the basis of safety studies (do not impose three barriers a priori, but rethink this notion, for the use of a solid fuel, according to this type of reactor).
- With regard to the risk of leakage from the fuel circuit by corrosion, develop provisions for prevention, detection, and limitation of the consequences.
- Analyses to be carried out on the entire installation (boiler, auxiliary circuits, salt treatment unit, etc.).
- Assessment of source terms and associated risks to be carried out.
- Evaluation of the composition and location of the radioactive inventory over time (including evacuation of part of the source term produced in the reactor).
- Activation of intermediate salt.
- Phenomenology in the event of fuel salt spillage in a room (solidification of salt/retention of cesium and iodine).
- Risk of reaction between fuel salt and other fluids.

Evacuation of residual power

- The fuel is liquid, which allows it to be relocated if necessary to ensure cooling.
- Objective of being able to manage a case of loss of the cooling circuits, including at the level of the emergency drain tank, without resorting to complex provisions, and taking into account the low thermal inertia of the salt.

- Special cases of filling/draining operations with a salt that can release residual heat and circulation in the intermediate exchanger that is not operational.
- Case of loss of cooling of the reactor structures to be assessed.
- Evacuation of the residual heat to be ensured throughout the installation (fuel salt, and also the bubbling and treatment unit), taking into account the evolution of the composition of the fuel.
- Other points to consider:
- Development and qualification of materials resistant to environmental constraints (temperature, irradiation, combustible salt, etc.).
- Development of continuous monitoring techniques (physical parameters and chemical composition).
- Limitation of worker exposure during operation and maintenance operations (dosimetric protection).
- Environmental impact.
- Chemical risks to be analyzed.
- Resistance to proliferation and physical protection.

Initiating event lists

A first list of initiating events (EIS) has been established [9], the main points of which are as follows:

- Reactivity insertion.
- Increased heat extraction/overcooling.
- Loss of fuel flow.
- Reduction of heat extraction.
- Loss of tightness of the fuel circuit.
- Loss of control over fuel salt composition/chemistry.
- Overheating of fuel circuit structures.
- Loss of cooling of other systems containing radioactive materials.
- Loss of containment of radioactive materials in other systems.
- Mechanical degradation of the fuel circuit.
- Loss of pressure control in the fuel circuit.
- Leak in the conversion circuit.
- Loss of power supply.

This list has made it possible to establish tighter lists of possible accidents and to begin studies of transients on certain increased accidental cases (e.g., insertion of +1000 pcm in 1 s, passage in 1 s from 100% to 0% from cooling to the intermediate exchanger, etc.).

Barriers

This analysis led to a first proposal of barriers for liquid fuel reactors:

- First barrier: structures containing the fuel circuit.
- Second barrier: recovery enclosure around the first barrier.
- Third barrier: reactor containment.

Proposed barriers for maintenance:

- First barrier: storage tanks + filling/drainage pipes.
- Second barrier: dedicated enclosure encompassing these structures.
- Third barrier: reactor building.

In the event of emptying into the emergency tank, this tank + fuel circuit can form a first barrier independent of the second and third barriers, but the integrity and tightness of this tank should then be maintained at the same level as the rest of the first barrier.

The current reflection on these barriers raises the problem of crossing these barriers, their minimization, and their management.

This reflection also makes it possible to expand the definition of a severe accident for an MSR, which could be the loss of tightness at the level of the second barrier, after a loss of the first. But a serious accident could also be the loss of the intermediate fluid with rise in temperature, melting of the exchanger walls, and flow of fuel toward the intermediate circuit. These proposals remain to be consolidated.

In these situations, the role of the second barrier is essentially to recover the salt/fuel.

Conclusion on safety

Studies show a number of positive points for the concept, including the passive safety provided by the high coefficients of feedback in the event of reactivity accidents. This aspect makes it possible to rethink the safety approach with a reduced maximum level of risk compared with the use of solid fuel (and with the resulting consequences in terms of acceptability).

On the other hand, a number of points linked in particular to the containment and evacuation of the residual heat will have to be studied in connection with the design of the reactor. They have thus far not been the subject of in-depth studies, and they will certainly have an influence on the design of the reactor.

Finally, the licensing of a reactor of this type must undergo approval by the safety authority on the basis of a number of definitions specific to MSRs, such as the definition of barriers, of the reference severe accident, of a safe stop situation without control rods, etc. It is in fact not possible, given the fundamental differences, to use the same framework to evaluate safety of solid fuel reactors for an MSR.

As it stands, a more precise reactor design is necessary. Once this is done, it will be possible to start designing a preliminary safety analysis, in particular with regard to conceivable accidents, and to produce results allowing an analysis by the safety authorities in the licensing process.

Waste and dismantling

We have seen that in operation a certain amount of waste should be accounted for, treated, and managed: gaseous purification, on-line or batch fuel treatment, and used components. The assessment of this waste and its management remains to be carried out, particularly with regard to the incineration options.

Dismantling has not been studied. The option of reusing the salt in the next reactor is suggested to limit the volume of fuel salt reclassified as final waste. This option, if validated, would simplify dismantling and significantly improve the final assessment of the waste produced.

For the dismantling itself, it should be recalled that solid fuel currently represents 99.9% of the residual activity of "conventional" reactors and much less than 1% of the final volume. Here, contamination by a liquid fuel will bring about a new and unprecedented situation, and it will be necessary to carry out a precise inventory of the volumes, but also of the installations and operations necessary for an MSR.

Reprocessing

The particularity of MSRs is that fuel reprocessing is carried out in parallel with operation. This is an overall advantage for the sector, as it can therefore considerably minimize the quantity of fuel outside the core. However, the design of the reactor must be developed simultaneously with its reprocessing mode, which makes its development more complex.

Thermal MSRs must reprocess fairly large quantities in parallel to avoid progressive poisoning of the reactor by fission products. Fast-spectrum MSRs require much lower quantities to be reprocessed (between 10 and 40 L/day for a 1400 MWe reactor) because there is less neutron poisoning by fission products in this spectrum.

Reprocessing also aims, in addition to the elimination of neutron poisons, to avoid the deposition of certain products (such as lanthanides) whose concentration increases over time and whose accumulation can disturb the solubility of actinides, in particular plutonium.

This reprocessing can be carried out by continuous sampling (on line) or discontinuously (by batch). The purified salt is returned to the fuel circuit with possible additions of products to be burned or incinerated, but also a certain quantity of fissile/fertile corresponding to the consumption of the reactor.

Note the indication in terms of size and volume, for on-line reprocessing: to give an order of magnitude, the average reprocessing speed would be around 1 L/h for a 1400 MWe reactor, which provides information on the space required and the volumes of the containers involved. Thus, there are batches of approximately 24 L/day.

France has high-performance installations for reprocessing by dissolution (PUREX process). These techniques are particularly interesting with chlorides, which dissolve very easily in water. Extraction methods exist to first recover the materials to be reinjected into the reactor, namely uranium, plutonium, and minor actinides. For the remaining products, the periods are short and the waste more easily manageable. It would be good to also recover the salt, which, owing to isotopic separation, has a cost that makes its recovery interesting.

It seems a priori more judicious to reprocess in a specialized central unit capable of both providing batches of new salts and taking back the batches of used salts, rather than building a reprocessing unit behind each reactor. The relatively small quantities and the possibility of a solidified salt during transport also militate for this solution, which is generally much simpler.

Proliferation

In terms of proliferation, plutonium, uranium-235, or uranium-233 are the fissile products to be controlled in these fuel cycles.

The existence of a central production and reprocessing unit, where these fissile flows are perfectly controllable, makes it possible to limit risk of proliferation.

The transport of solidified batches between this central unit and the reactor makes it possible to minimize the risks because the quantities of fissile transported remain low and their extraction would require complex technologies.

However, the main point is that the plutonium of MSR fuel is obtained by reprocessing of PWR used fuel. And as explained in Chapter 4, this isotopic composition does not allow any military use.

Conclusion

Molten salt reactors have extremely interesting potential advantages vis-à-vis SFRs. But, unlike the SFRs, for which the feedback is very high, there is little feedback available.

This has two consequences. The first is that major developments are to be carried out, particularly the construction of demonstration reactors joined, thanks to code developments, to digital twins. The second is that unknowns concerning the feasibility and reliability of these reactors remain until these developments have been carried out.

We also know that concepts that appear simple at the start often become more complex as problems appear.

A final comparison between the two concepts of fast reactors can be made only after these necessary developments on the MSRs.

References

[1] MIT report of December 2014; "Fluoride-Salt-Cooled High-Temperature Reactor (FHR) for Power and Process Heat" Final Project Report Charles Forsberg, Lin-Wen Hu, Per Peterson, Kumar Sridharan.

[2] MSRE documentation (Oak Ridge). http://energyfromthorium.com/pdf/.

[3] Documentation on the MSFR. http://lpsc.in2p3.fr/index.php/fr/groupes-de-physique/implications-societaux/msfr/rsf-reacteurs-a-sels-fondus/lang-fr-msfr-bibliography-lang-lang-en-msfr-bibliography-lang.

[4] ICAPP 2019 Molten Salt Reactor to close the fuel cycle: example of MSFR multi-recycling application J. Guidez and col....

[5] Article for nuclear sciences and techniques October 2019 "Influences of 37Cl enrichment on neutron physics in Molten Chloride Salt Fast Reactor_" Liao-Yuan He, and col. University of Chinese Academy of Sciences, Beijing, China 100049.

[6] R. Cury, Metallurgical Study of Ni-W and Ni-W-CR Alloys: Relationship Between Short-range Order and Hardening (thesis), University of Paris XII, 2007.

[7] ORNL/ER-341. Program management plan for the Molten Salt Reactor Experiment remediation project at Oakridge National Laboratory.
[8] STJ-02MSRE-D992—Molten Salt Reactor Experiment Engineering Evaluation and Extended Life Study, URS | CH2M Oak Ridge LLC, Oak Ridge, Tennessee.
[9] Delphine Gérardin's thesis, Development of Digital Tools and Performance of Studies for the Management and Safety of the MSFR Molten Salt Reactor, Grenoble-Alpes University, 2017.

CHAPTER

Conclusion: What future for fast reactors and our planet?

8

Energy availability is a key element for any human civilization and its development. Energy savings are possible in many areas, but experience shows that the results are limited, unless certain uses are eliminated.

All the available projections show an increase in global energy demand in the short term, due to demography, the necessary catching up of billions of people who have insufficient access to energy, and new uses such as digital technology.

Fossil fuels have allowed our civilizations to take off economically, but their golden age is ending for several reasons. For one, certain resources are being depleted. The production peak of crude oil was reached in 2018. Currently, production is carried out by means that are increasingly expensive and polluting: hydraulic fracture, tar sands, etc. For other fossil fuels, gas and especially coal, there are still hundreds of years of production left, but here too the prices and pollution will continue to increase. However, the main reason we should accelerate the inevitable phasing out of these fossil fuels is the climatic upheaval that these energies are causing by the release of very large amounts of CO_2 and the corresponding greenhouse effect.

The development of renewable energies is useful but everyone knows that they will not be enough, in particular because they are not controllable and we do not know how to store electricity in large quantities. In addition, the ecological consequences of a massive deployment are not negligible, in particular in terms of footprint. Their development must therefore be accompanied by the development of controllable nuclear energies in order to try together to replace fossil fuels which currently cover more than 85% of the planet's energy needs.

The current second- and third-generation reactors use the uranium resource, which is not inexhaustible and on which speculations have started. They use enriched uranium and store depleted uranium. Then they produce spent fuel or long-lived waste after reprocessing.

In this context, fast reactors make it possible to provide a necessary and indispensable complement because they allow the development of nuclear power using these existing wastes created by second- and third-generation reactors, that is to say depleted uranium and the products issued from reprocessing. They do not need uranium mines and enrichment factories anymore. They only produce short-lived, easily manageable waste.

Among the range of fourth-generation reactors, only two types of reactors today can claim to provide these solutions as fast reactors.

Fast Reactors. https://doi.org/10.1016/B978-0-12-821946-1.00009-7

The best known is the sodium-cooled fast reactor (SFR). Many reactors of this type have been built, and a great amount of experience feedback exists for it, making it possible to improve its design, in particular with regard to safety and simplicity. The latest design of this reactor is presented in Chapter 6 with the ESFR-SMART project. An update on the fuels available for this reactor is given in Chapter 5.

However, an analysis of why the deployment of these reactors is so slow, and why blockage occurs, is carried out in Chapter 4. The analysis reveals that this sector, in addition to reactors, must also develop reprocessing and fuel fabrication plants, which represent substantial investments. The cost of the reactors alone is higher than the cost of current water reactors. The interest is therefore not economic, as long as uranium prices remain low, but ecological if one seeks to close the cycle and minimize waste.

Another type of fast reactor with great potential is the molten salt reactor (MSR). Its analysis, carried out in Chapter 7, reveals its significant advantages in terms of safety, cost, fuel cycle, waste incineration, and flexibility of use. Unfortunately, experience feedback on this type of reactor is low, and many developments are needed to confirm this high potential. Our technical knowledge of these reactors is improving, and many projects are in progress, in particular the construction of demonstrators. It is clear that, if these demonstrators confirmed the potential advantages, this type of reactor would also present a great advantage in terms of acceptability to society.

What will become of our planet? Will there really be a gradual abandonment of fossil fuels in favor of carbon-free energies? What role will nuclear energy play globally in the years to come? And what role will there be for fast reactors?

On the other hand, without political consideration and on a strictly technical level, it should be clear to the reader, at the time of this book's release, that humanity has, with these SFR and MSR fast reactors, technical solutions capable of providing it with energy in the millennia to come, without producing CO_2 and using only already existing waste.

Appendix 1: Lessons learned from sodium-cooled fast reactor operation

The purpose of this appendix is to show that, thanks to the significant feedback from sodium reactors around the world, most of the teething problems of these reactors have now been resolved. In this appendix, we provide solutions for each problem.

It should also be noted that cooperation on the subject was very strong until the 2000s, in particular through meetings held periodically by the International Atomic Energy Agency, where the problems that arose were discussed and exchanged. More recently, serious problems have arisen on these reactors without being the subject of publications or exchanges in the international community. At the start of BN-800, there were serious problems with the assemblies, resulting in a 1-year delay in the schedule. The filling of the sodium loops of the Indian reactor has been going on for 4 years without an explanation for the delay, and the China Experimental Fast Reactor (CEFR) reactor has had serious air intake problems. The author will address all these events, but if he has gained information through private discussions during meetings, no publication reference can support these explanations.

Therefore, the overview of these problems and their respective solution cannot be completely exhaustive.

Sodium-water reactions

All the first-generation steam generators experienced sodium-water reactions, and this problem remains in the mind as a major problem despite having in fact been completely solved today.

- Phénix: five sodium-water reactions (Ref. [1], Chapter 9)
- BN-350: several reactions, one very big reaction
- Prototype Fast Reactor (PFR): 21 reactions, one very important reaction
- BN-600: several reactions at the beginning of operation

Superphénix [2] and Fast Breeder Test Reactor (FBTR), however, are exceptions, without any sodium-water reaction, and BN-800 also since the beginning of its operation.

Although the sodium-water reactions were identified in different Steam Generator (SG) designs, three types of causes can be defined.

- Manufacturing problems
  Several sodium-water reactions occurred at sodium filling or shortly afterwards, owing to constructional faults, in particular in the Russian reactors. In the PFR reactor, a crack was found in a sheet metal/plate weld during filling (1976).
- Fatigue cracks
  The combination of a design flaw and inappropriate operating procedures led to thermal shocks and mechanical fatigue. Fatigue cracks can initiate sodium-water reactions. This was the case for the first four reactions at Phénix, where the spurious arrival of cold water in the reheater during startup caused fatigue cracking.
- Corrosion
  Corrosion phenomena can lead to a surface-breaking crack and a sodium-water reaction. This was the case for the PFR, where repeated corrosion occurred at the tube-plate junctions. This was also the case for the fifth sodium-water reaction in Phénix.

In all cases, the pressurized steam jet increased rapidly. In reacting with the sodium, it created a torch whose inner cone can quickly bore through the nearby steam tubes of the SG shell (wastage effect) and rapidly increase the damage. In the first prototype reactors, these early sodium-water reactions caused significant damage due to the too long response time before detection and the time to drain the steam generator.

Some examples of sodium-water reactions:

- The reaction that occurred in February 1987 at PFR, in the main section of the superheater tube, where a tube cracked following bundle vibrations. Holes were bored into the 39 tubes around the initial tube during the time it took to lower the steam pressure. The corresponding rise in pressure in the secondary sodium circuit caused the burst diaphragm to rupture.
- Significant sodium-water reactions occurred in BN-350 and on BN-600 (the biggest event took place in January 1981 with 40 kg of water injected).

The following lessons were learned from these incidents, for incorporation in future reactor design:

In terms of protection:

- Reliability and rapidity of the hydrogen detection system.
- Automatic shutdown accompanied by rapid depressurization on the steam side.
- Design of a casing around the self-supporting SG, capable of confining even the most violent sodium-water reaction.

In terms of prevention, there are several requirements to avoid particular problems in future reactors:

- Better thermomechanical design.
- Right materials.

- Extreme quality in manufacturing (100% inspection, nonextended welds, etc.).
- Precautions in use (prevention of spurious thermal shocks, circuit well protected during drainage to avoid corrosion, etc.).

It should be pointed out that Superphénix, which benefited from this feedback, did not have any sodium-water reactions during the 10 years of operation. Likewise, since 1991, BN-600 has had no further sodium-water reactions, nor has BN-800 since its startup.

Handling operations

Blind handling of the core elements in sodium was performed well overall for all the reactors.

A serious incident was a spurious rotation of the core cover plug during a subassembly handling in FBTR in 1987, which led to deformation of a fuel subassembly and a guide tube. Manufacture of special devices created to shear them, then remove them, led to a 2-year reactor shutdown.

The same incident occurred in JOYO in 2008, with a spurious rotation of the cover plug with an experiment in the core. The damage caused in this plug led to its replacement 3 years later.

In these two cases, the use of an ultrasound vision system, as in Phénix, capable of evaluating the risk of plug rotation before it occurs would have avoided these damages.

In conclusion, when using an effective core cover plug ultrasound vision system, sodium handling operates satisfactorily.

Operation of primary components (pumps and exchangers)

These components displayed excellent operation.

The following observations are made:

- For the Phénix exchangers, an initial design defect that was corrected in Phénix and all following rapid reactors (Ref. [1], Chapter 8).
- For the pumps, a technological defect on the hydraulic bearing of the Phénix pumps (bearing ringed on the shaft, falling during a hot thermal shock; Ref. [1], Chapter 10) and a filter problem in a PFR pump during a major oil loss in primary sodium [3].

In conclusion, these components operate well and should pose no particular problem in future reactors.

Spurious leaks or transfers of sodium

Several sodium leaks took place on the installations, sometimes leading to sodium fires. The most serious sodium leak, in terms of the consequences, occurred at MONJU (~ 640 kg of Na), which resulted in a 12-year reactor shutdown.

These leaks involved very different volumes of sodium. Some involved less than 1 g (detected during inspection), while others were massive (BN-600, two leaks > 300 kg and one involving 1000 kg) [3,4].

These sodium leaks can have many very different origins:

- Constructional defects
- Design problems, such as the MONJU thermocouple thimble
- Materials problems, such as the example of type 321 steel stress cracking
- Thermal striping at the mixing tee level, leading to through cracks
- Corrosion following air intake into the circuits
- Operator error (e.g., during heating of the circuit and the corresponding expansion of the sodium)

Lessons have been learned from these incidents in terms of design, circuit operating procedures, leak detection, and protection from sodium fires.

For the design of the reactor, we must take into account the following points:

- Diversified, redundant detection instrumentation (bead detector, sodium aerosol detector, smoke detector, camera, etc.)
- Need for rapid drainage of the sodium circuits.
- Sectoring around the sodium areas to limit the quantity of air available in the event of a possible fire.
- Insulation protection of the concrete floors and walls, covered with a metal plate.

In Phénix, we have 31 sodium leaks (Ref. [1], Chapter 20). In Superphénix, almost all these leaks are suppressed (Ref. [2], Chapter 14). In ESFR-SMART with straight pipes in the secondary loops, we could continue to improve prevention and detection.

Intake of air or impurities or gas

In a sodium reactor, avoiding the intake of air or impurities into the circuits is of utmost importance. Under certain conditions, these pollutions can start mechanisms of stress-corrosion cracking. Outside of the many ongoing operational problems that this constraint entails, particularly for conservation of the circuits during drainage, there were several significant incidents in this area:

- Superphénix
  Significant air intake occurred in July 1990, in the primary circuit of Superphénix, due to a defective membrane on a compressor sending air flow to the reactor cover gas. A major purification campaign, through to September 1991, was required to restart the reactor (Ref. [2], Chapter 5).
- PFR
  Significant oil intake (~17 L) occurred in June 1991. This resulted in partial clogging of the pump filter, and was detected by temperature variations at the core outlet. An 18-month shutdown was required to recondition the reactor. Lessons have been learned from these incidents in terms of design and circuit operation.

- BN-600
  In 1987, the BN-600 reactor underwent a brief transient at rated power. On January 21, during 7 min, with the reactor at rated power, changes were noted in several parameters, including sodium level, core reactivity, and reduced power at the pumps. After analysis, the incident was attributed to the drop of a chunk of impurities from the reactor roof, formed at the top by contributions from poorly purified gas. These impurities, which fell suddenly in a block, disturbed core hydraulics and neutronics. This event then later caused several clad failures and reactor shutdown. Since this time, gas leaks in the core cover gas have been minimized, and more advanced purification of the argon has taken place. This phenomenon has not reoccurred since 1987.
- CEFR
  In this experimental Chinese reactor, air intake occurred in the year 2000 during tentative work to resolve handling problems. This stopped the reactor operation for a long period. However, no publication is available on the subject. Feedback from such incidents has of course revealed the need for proper surveillance of the quality of the inputs from core cover gas, and a pump design that precludes oil drops (intermediary recovery).
  Gas entrainment is also a design problem. Many propositions were made on Astrid and ESFR-SMART projects to avoid these problems in the future.
  We suspect that, during the filling of the secondary loop of the PBFR, it was revealed during operation that there was significant entrainment of gas by vortices in the expansion tank of the secondary pump. This design problem could be the cause of the startup delay of this reactor. However, no publication confirms this scenario in 2022.

Experience from fuel and clad failures

Most of the reactors used uranium oxide and plutonium fuel, with excellent feedback experience. However other fuels were used, such as enriched uranium (BN-600), carbide fuel (FBTR), and metallic fuels for the American reactors. For all these fuels, the progress made on clad materials has gradually led to eliminating clad failures and significantly increasing burnup.

On the subject of the clad failures on the German KNK reactor (Kompakte Natriumgekühlte Kernreaktoranlage), these were due to the use of grids instead of spacer wires. Since that time, all fast reactors have used spacer wires to separate the pins.

In Phénix, after several clad failures, the problem was resolved (Ref. [1], Chapter 7), and no clad failure occurred on Superphénix. The same is true for BN-600, then BN-800.

In conclusion, this field is a strong point for this reactor type, and research and development seeks to make further progress in this field through the use of new cladding materials.

Material problems

The prototype aspect of these reactors served to validate the use of many materials, and to continually improve these materials. One example is the move to ferritic steel for fuel subassembly wrapper tubes, which eliminated the swelling in the subassemblies. For example, the BOITIX experiment at Phénix reached a dose of 155 dpa (Ref. [1], Chapter 7).

The two biggest problems encountered in this area were the following:

- Extensive use of type 321 austenitic steel in Phénix and PFR. This steel developed cracks over time corresponding to residual welding stresses, particularly in the hot areas. As a result, all the type 321 parts in Phénix were gradually replaced. Many successive repairs were made to the PFR steam generators, and all the parts composed of type 321 on existing reactors are closely monitored (Ref. [1], Chapter 16).
- The Superphénix drum cracking led to the abandonment of this material (15 D3 steel) for use with sodium (Ref. [2], Chapter 17).

Today, there is a range of approved materials. However, development of new materials remains one of the promising directions for continuous improvement for this reactor type. In particular, oxide-dispersion steels could enable one to reach a dose of 200 dpa for the clads. The secondary circuit materials, with reduced expansion coefficient, allow the use of straight tubes as proposed on ESFR-SMART secondary circuits (see Chapter 6 of this book).

The qualification of these new materials remains to be achieved.

Neutronic operations and control

Several characteristics intrinsic to the sodium-cooled fast reactor type result in straightforward reactor operations and wide safety margins:

- Self-stabilizing thermal counter-reactions
- The absence of any fission product poisoning phenomenon
- No use of neutronic poison for reactivity control
- High thermal inertia from the volume of primary sodium
- Very low fluid pressure

There are no particular problems or difficulties to report from the operations of any of the reactors, which are very easy to operate (compared with water reactors).

However, core sensitivity to reactivity variations, during movement of the subassemblies, should be pointed out.

In FBTR, this led to positive reactivity variations through repositioning of the core at reduced power. These events have been explained and are reproducible.

In Phénix, core compacting is impossible (subassemblies are in contact at the level of the subassembly wrapper tube plates during reactor operation). However, radial outward movement is mechanically possible and can lead to negative reactivity variations. Such reactivity transients occurred at Phénix in 1989 and 1990. The investigations into the mechanism leading to the radial outward movement have developed several explanatory scenarios. The most likely scenario is based on the overheating of sodium in an experimental device near the core (see Ref. [1] in Chapter 24).

A strange and new incident occurred during the startup of the BN-800 reactor. A change in reactivity occurred and was attributed to the untimely rise of an assembly. On analysis, it appeared that a defective design of the assembly feet had led to vibrations and a separation of the assembly with its shrunk and not welded foot. This led to a 1-year delay in starting the reactor, to allow modification of all these assembly feet and their method of attachment. Again, no publication is available on the subject.

Sodium aerosols

Convection movement of the reactor cover gas from the lower sodium hot areas near the sodium surface to the colder zones carries sodium aerosols that can be deposited on the upper parts of the reactor. This experience, applied to the design of future reactors, represents significant progress toward reaching the objectives of good performance, competitiveness, and reliability assigned to this reactor type. This phenomenon has led to several operating constraints. Globally, reactors have experienced problems related to deposits of sodium aerosols. Two particular events are cases in point:

- In BN-600 in 1997, significantly greater efforts were required to move the rotating plug. Analysis showed large deposits of aerosols on the bearings (and in the bearing cage). The system was disassembled and cleaned [3,4].
- In several reactors (PFR, KNK II, Phénix, etc.), aerosol deposits led to partial blocking of the control rod. These incidents led to design changes and improvement of the monitoring instrumentation and operating procedures, including, among others, periodic verification tests of the control rods performed on the reactor in operation (Ref. [1], Chapter 11).

Conclusion

Sodium-cooled fast reactors have now built up significant feedback experience in the fields of materials, design, sodium technology, and operating modes. Analysis of reactor availability factors has shown that, in the past, availability has been affected by the difficulties inherent in the role as prototypes. However, once this burn-in work on initial design errors has been accomplished, these reactors show outstanding ease of use and reliability.

Operations with BN-600, once steam generator problems and early fuel behavior problems were solved, are an excellent case in point. Since 1990, the BN 600 has continuously displayed load factors around the 80% level.

Similarly, Phénix has made significant contributions to this reactor type, by successively modifying component design (heat exchangers, pumps, etc.), operating procedures, and core materials. The reactor's average availability factor over the first 17 years of operation was 60%. Since the reactor started back up in 2003, after the renovation phase, availability factor has been about 75%. Production losses have been nearly exclusively attributed to the classical part of the reactor (electricity production installation) and not to the sodium part.

A reactor design, taking into account all this feedback experience, should now reach a very good availability factor.

References

[1] M. Guidez, Phenix: The Experience Feedback, Edited by EDP Sciences, 2013.

[2] M.M. Guidez, G. Prele, Superphenix: Technical and Scientific Achievements, Edited by Springer Editions, 2017.

[3] J. Guidez, et al., Lessons learned from sodium cooled fast reactor operation and their ramifications for future reactors with respect to enhanced safety and reliability, Nucl Technol (November 2008).

[4] J. Guidez, L. Martin, Review of the experience with worldwide sodium fast reactor operation and application to future design, in: IAEA International Conference on Research Reactors, Sydney (Australia), November 2007.

Appendix 2: Industrial demonstration of the fuel cycle of a sodium-cooled fast reactor

As explained in Chapter 2 of this book, the advantages of a fast reactor lie in its fuel cycle:

- Manufacture of fuel with "waste" from the cycle of existing reactors, that is, depleted uranium from enrichment plants, and plutonium from fuel reprocessing.
- Use of fuel in the reactor.
- Reprocessing of spent fuel with recovery of uranium and plutonium, or even other minor actinides. Final disposal of the remaining products, which are "short-lived" products.
- Reuse of uranium and plutonium to manufacture new fuels.

This cycle has been demonstrated on a laboratory scale (a few hundred kilos) for the American EBR2 reactor and in Russia where a modern fuel fabrication plant has been inaugurated (see Chapter 5).

But the most significant industrial experience today remains the experience of the Phénix reactor, which dates to the 1980s.

All this experience, which is without equivalent in the world, on the reprocessing of fast fuel with a high burnup rate, on the vitrification of the corresponding waste, and on the closure of the cycle with several successive passages has demonstrated industrially the feasibility of the fast reactor fuel cycle.

Phénix fuel reprocessing experiment

- General data:
  520 Phénix assemblies were reprocessed from 1976 to 1993, which corresponds to the extraction of 4.4 tons of plutonium metal.
  These reprocessing took place successively in three facilities:
- The experimental pilot workshop for the reprocessing of FNR fuels: AT1 workshop at La Hague, from 1969 to 1977. The maximum capacity was 1 kg/day (150 kg/year).
- The Marcoule pilot workshop (APM) from 1973 to 1997, the capacity of which increased from 2 to 5 tons/year.
- The UP2-400 plant at La Hague where reprocessing is carried out in dilution with *Uranium Naturel Graphite Gaz* (UNGG) fuels.

Review of experience feedback on reprocessing techniques

- Dismantling of assemblies
  The very thick hexagonal tube of fast reactor assemblies prohibits shearing of the entire assembly, as practiced for water reactors. Assemblies stripped of their sodium must therefore be dismantled to extract the pins. This operation was carried out in the examination cell (CE) of the Phénix plant, by mechanical milling on the angle, then spreading and extracting bundles of pins placed in cases.
- Pin shearing
  For this operation, the question arises of whether it is necessary to remove the spacer wire beforehand. The answer depends on how the shear and dissolver work. It should be noted that the Phénix experiment was carried out without prior removal of the wires. The shearing in sections was carried out without problems, pin by pin (AT1), on rows of pins (TOR), or by direct shearing of the cases (UP2-400).
- Fuel dissolution
  Two aspects need to be considered: bringing the uranium and plutonium into solution as completely as possible in boiling nitric acid, and the resistance of the cladding material in this environment.
  Uranium oxide and plutonium oxide form in nitric acid solid solutions up to a certain ratio (about 35% plutonium) beyond which they behave as insolubles. The Phénix cores (18% plutonium in core 1 and 25% in core 2) did not cause any problems within the range of burn rates tested.
  The cladding materials tested, 316L, 316Ti, and 15-15Ti, showed good behavior.
- Clarification
  The separation of insoluble products before extraction operations does not pose a problem specific to fast fuel.
- Extraction cycles
  These operations are not specific to fast fuels either, and the improvements made in the context of water reactors can be transposed.
  It should be noted, however, that it is necessary to cool the, assemblies for at least 5 years, for the solutions to remain within the usual standards of thermal power.
- Waste
  There are few specific problems. We can note the validation of the vitrification of fission products. For the specific problem of the stainless steel hexagonal tubes, fusion seems to be the best development process, and has been tested on several kilograms.

Conclusion on the reprocessing

In total, 520 assemblies, or the equivalent of approximately 4.5 Phénix cores, were reprocessed. This represents, taking into account the 2.3 tons of the first cores enriched in UO_2, a little more than 26 tons.

It should also be noted that the measurements and balances carried out during the reprocessing operations made it possible to measure an overall breeding rate of 1.16 and thus confirm experimentally the expected theoretical values (1.13).

Feedback on fuel fabrication

All Phénix fuels were manufactured in the ATPu facility in Cadarache, which also manufactured Rapsodie and Superphénix fuels.

An important objective was to reuse this plutonium produced by reprocessing to manufacture new Phénix assemblies and to show industrially the possibilities of multirecycling of fast reactors.

Of the 4.4 tons of plutonium produced, 3.3 tons were reused to generate fuel for Phénix.

In 1980, the first assembly made with recycled plutonium from Phénix was loaded into the reactor and carried out the first cycle closure.

This experiment culminated with the CPD 408 assembly (Fontenoy Experiment), which used recycled plutonium for the third time.

The plutonium obtained at the APM was sent to La Hague, in the form of nitrate, for conversion into oxide. As this conversion was carried out with dilution in UNGG fuel products, it was not possible to monitor the isotopic evolution of the fuel as it passed through the reactor.

This isotopic evolution was linked, historically, to the origin of the Pu used: UNGG, Phénix, then REL, as well as to its aging, which in particular induces a rise in americium levels.

Applications for the reactors of the future

Breeding rates and reprocessing possibilities have been industrially demonstrated thanks to the reprocessing of these 520 Phénix assemblies.

To manufacture new Phénix assemblies, 3.3 tons of the 4.4 tons of plutonium extracted were reused, by reburning in the reactor.

For future reactors, it should be noted that increases in Pu content, increases in burnup, or changes in clad material require validations at the level of the dissolution process.

Apart from these reservations, and notwithstanding the improvements that are always possible, the feasibility of reprocessing and industrial fabrication of fuels with the plutonium resulting from this reprocessing has been demonstrated industrially.

Similarly, an industrial demonstration was provided of the feasibility of fuel fabrication, using this reprocessed plutonium, and the possibility of its recycling in the reactor.

Further reading

M.J. Guidez, Chapter 25: Reprocessing and multirecycling, in: Phenix: The Experience Feedback, EDP Sciences, 2013.

Appendix 3: Choice of fuel salt for a U/Pu cycle molten salt reactor

In a molten salt reactor (MSR), it is necessary to dissolve in salt the fissile elements (which trigger the desired nuclear reaction) and fertile elements (which make it possible to re-create fissile elements and thus remain in the virtuous circle of the cycle of fast reactor fuels).

For the fissile, we have the choice between enriched uranium, plutonium from the reprocessing of spent fuel, or even other minor actinides from this same reprocessing (such as americium). The choice depends primarily on the capabilities of the country where the reactor is developed. In France, the uranium enrichment plant supplies only low-enriched uranium for water reactors. An MSR would require highly enriched uranium, that is, 20%, the international limit accepted for nonproliferation reasons. On the other hand, the reprocessing of fuel from water reactors produces around 10 tons of plutonium per year. Currently, as this same reprocessing does not produce the other minor actinides (such as americium), they are not available today. The simplest and most pragmatic choice is therefore to use plutonium as fissile and to recycle waste. This does not exclude the use in the longer term (and certainly after using all the stock of plutonium available) of uranium enriched to 20% (but a dedicated enrichment plant would need to be built) or americium (but the current reprocessing chains would need to be modified).

For the fertile, we can use depleted uranium from enrichment plants, or thorium from mining activities. The availability of these two products is not a problem. Uranium will produce plutonium during operation and thus make the reactor isogenerating or even breeder-generating. The thorium produces as fissile uranium-233, which is an excellent fissile. The extraction of this uranium-233 would make it possible in the long term to create a thorium/^{233}U cycle that would no longer require 20% enriched uranium.

For salts, the choice is very vast. It is recalled that the only operational feedback we have is for the thermal flux Molten-Salt Reactor Experiment (MSRE) feedback, with 33% enriched uranium-235 as the fissile. This enrichment, the thermal flux, and the short operating time led to a low final production of Pu (about 600 g). A second operating cycle was also carried out with uranium-233 as the fissile, but thorium was never used. In both cases, the salts used were fluoride salts. No reprocessing of salts was carried out.

In the context of a fast reactor with U/Pu cycle, the salt selection parameters are different. The purpose of this appendix is to gather certain data in order to be able to propose a choice of fuel salt for an MSR with fast spectrum and U/Pu cycle, that is, a fast reactor using the products of the cycle available in France: depleted U (more than 300,000 tons), and plutonium from the reprocessing of pressurized water reactor (PWR) and mixed oxide (MOX) fuel [1].

The objectives of this rapid MSR are therefore:

(1) Use the existing products available from the French cycle (depleted U, Pu and products from the reprocessing of PWR fuels, or spent MOX).
(2) Transmute minor actinides in case of subsequent separation during reprocessing.
(3) Minimize the activity of the end products of the cycle.
(4) Facilitate management of plutonium stocks by using burner, isogenerator, or fast breeder modes.
(5) Ensure reactor has elements of intrinsic safety

The ideal salt (which does not exist!) should meet the following parameters:

(1) Criticality, operating, and melting temperatures
The melting point of pure salt is a characteristic that needs to be considered. However, the introduction of U and Pu leads to different, generally higher, eutectic temperatures. The criticality temperature is the average operating temperature at which the reactor is critical with its U and Pu composition, and in the core geometry.
For example, with a criticality temperature of 700°C and a core temperature of 90°C, the fuel salt temperature varies between 655°C at the core inlet and 745°C at the outlet.
Accidents during operation, hot shock, and cold shock will extend this operating temperature range.
It should be noted that, considering the resistance of the materials available today, it would be necessary to be able to operate at maximum temperatures below 700°C.
(2) Solubility of U, Pu, and minor actinides
The solubility of U, Pu, etc., must be sufficient in the selected salt to reach criticality, necessary for operation, at the criticality temperature.
However, this solubility generally decreases with temperature. However, it is necessary that, over the entire range between melting temperature and criticality temperature, this solubility remains sufficient to avoid risk of precipitation and therefore of a material concentration at one point of the reactor causing a possible reactivity incident (for example, in the case of Pu). To meet our intrinsic safety criteria, the salt should solidify before the formation of a precipitate.
(3) Salt reprocessing
The objective of salt (re)treatment is to maintain the physical and chemical characteristics of the salt necessary for continued operation (fluidity, solubility rate, composition, behavior of materials, monitoring of redox potential, etc.).
This processing phase also makes it possible to maintain the neutron characteristics: recovery of U, Pu, and minor actinides to put them back into the fuel salt, elimination of certain unstable or stable fission products (e.g., zirconium), and, of course, addition of fissile, fertile, or minor actinide elements according to criticality needs.

Finally, salt must also be recovered, first of all because it constitutes an expense, and above all to limit the quantity of final waste. Everything else is final waste. Quantities: about 40 L/day of fuel salt for a 3 GWth MSFR type reactor with fast spectrum and U/Pu cycle. We reach ~4000 L/day in case of thermal spectrum. The reprocessing processes are different depending on the choice of salt and will therefore influence this choice.

(4) Corrosion/materials
The fuel salt must be compatible with the materials available.
Today, with the feedback experience of MSRE, Hastelloy (and its variants) is qualified for operation at up to 700°C with fluoride salts.
In fact, corrosion is not due to the salt itself, but to the various products created during operation. These products can be extremely diverse, and the redox potential must be controlled to avoid these corrosions. Indeed, the fission products formed by the fission reaction are produced at the redox potential of the salt (they cannot be more oxidizing than the salt). However, some fission products may have complexing action, which increases the tendency of the material to oxidize. Finally, some products (such as tellurium and sulfur) can also cause other types of corrosion, but here again, in principle, redox control should limit corrosion.
We see that, in this context, corrosion is not a criterion for choosing the salt, but rather a reason for research on the mode of protection for a given salt. A key point in the design of MSRs is the measurement and control of the redox potential of the fuel salt.

(5) Stability of the eutectic salt
One way to lower the operating temperature of the salt is to use a eutectic. Its stability must be demonstrated in the operating range, and its solidification behavior must be known (leak, etc.).
In this context, FliBe or FliNak have, for example, been proposed to operate at a lower melting temperature than pure LiF. However, the addition of large quantities of solutes (thorium, U, Pu, etc.) will modify this melting temperature by creating a new eutectic.
Another point is the variation of the melting temperature of the salt according to the variation of its composition. It is clear that the lower a salt's variation in temperature in the event of a variation in its composition, the easier it is to control, offering wider operating margins and, therefore, higher reliability.

(6) Neutron properties
We are in rapid flux. Depending on the salt retained, the captures and moderation of neutrons are more or less important, as well as the displacement per atom reached on the different materials. Finally, if certain elements of the solvent absorb too many neutrons, criticality or regeneration may no longer be achievable.

(7) Resistance to irradiation/formation of troublesome products
Some constituents of salts can form troublesome activation products (Cl-36, tritium, etc.); others are neutron poisons (Li-6, Zr). This leads to the need for prior isotopic separations on certain salts: elimination of Li-6 for lithium fluoride salts and of Cl-35 for chlorides.

(8) Feedback experience available for nuclear field
Some salts have been more studied and tested, and have an interesting feedback experience.
Some salts were used in reactor (including FLiBe used in the MSRE reactor), and others for (re)treatment (such as chlorides).

(9) Feedback experience available in the industry
Some salts have been or are used in industry, providing a source of tools and technical knowledge or technologies.
For example, if isotopic separation necessary, using techniques that already exist will save a lot of research and development (R&D) time.

(10) Reaction with air and water
It is of interest to avoid chemical reactions, especially in accidental situations, with air and with water. A physical reaction with high-temperature liquid salt and liquid water is unavoidable.
Some salts can be dissolved by water (chloride salts) while others cannot (fluoride salts), which introduces advantages/disadvantages depending on the situation considered.

(11) Chemical toxicity
Certain products such as beryllium are very penalizing in terms of toxicity. Their use is very difficult (even prohibited in some installations) with existing French regulations. They should therefore be avoided if possible.

(12) Physical properties: heat capacity, conductivity, viscosity
Good electrical conductivity would allow, for example, the use of electromagnetic pumps. A high heat capacity makes it possible to reduce the flow rates. The vapor pressure must not be too low (to avoid inducing evaporation of the salt when the temperature is too high). A high viscosity would be penalizing in terms of pressure drops.

(13) Salt control in operation
It is necessary to be able to control the salt according to different aspects, for example, in terms of corrosion (redox) and composition. This aspect is highly dependent on existing techniques.

(14) Simplicity/difficulties of implementation for operation and maintenance
It is necessary to have a salt that can be easily implemented in an installation. Indeed, a salt can have the right chemical and neutronic properties but prove to be difficult to handle owing to its physical characteristics and, thus, make the operation of the reactor complex and difficult.

(15) Level of technical mastery of salt
It is better to have a salt for which there are already objects, materials, and knowledge on its technical implementation available. A salt for which there are few elements will require longer R&D.

(16) Availability and cost
The desired salts must be available and affordable. If isotopic separations are necessary, their cost must be taken into account.

It should be noted that these parameters do not have equal weight. Some will prove to be prohibitive for a given choice, while for the most part we will have to be satisfied with the results obtained even if they are not optimal.

It should also be noted that the salt must be chosen on the basis of the overall design integrating the (re)treatment of the salt, which complicates the choice compared with the salts proposed as simple cooling fluids in solid fuel reactors.

In fact, it seems that three essential parameters will determine the choice:

(1) The melting and operating temperature of the initial mixture.
(2) Sufficient solubility in salt of U and Pu to reach criticality, while avoiding untimely depositions, in intermediate temperature ranges.
(3) The possibilities of processing the retained salt.

We will not find a miracle fuel salt that will respond optimally to all the parameters mentioned. If we find a salt that meets the three criteria above and meets a majority of other criteria, then we will have to accept a certain number of corresponding but identified defects.

There are two large families of possible salts: chlorides and fluorides, with an almost infinite variety of choices depending on the possible additions (Be, Na, K, etc.).

Chlorides:

They have some advantages: lower operating temperature, harder fast spectrum, cost and availability, good solubility of U and Pu.

We will also see that their reprocessing appears simpler.

On the other hand, they have a number of drawbacks: lack of feedback experience in the MSR reactor, higher level of irradiation on the structures than with fluoride salts, need for isotopic separation (to avoid chlorine-35, which forms chlorine-36, a neutron poison).

Finally, they have the characteristic of binding with the ambient humidity in certain temperature ranges, which can be annoying in the event of a leak, but can represent an advantage for the maintenance of the reactor because it allows the cleaning of the components.

The salts studied in the various references are generally based on NaCl combined or not with another salt.

Startups aiming for fast fluxes to burn waste (TerraPower and Elysium's MCSFR) have chosen to use NaCl-based chloride salts.

Fluorides:

They are widely studied as fuel salt, because of the feedback experience of the MSRE.

They lead to a higher operating temperature and have lower U and Pu dissolution rates than chloride salts. As fluorides moderate neutrons more, the spectrum is slightly slower (even if it remains fast), but this effect is sufficient to reduce the irradiation on the structures. Furthermore, the lower ionic radius of fluorides has the consequence of allowing smaller volumes for critical areas.

They are not soluble in water, which is an advantage in the event of a combustible salt leak because the radioactive elements remain trapped in the solidified salt.

However, this can pose complications for the maintenance and operation of a reactor by making it impossible to clean components or structures with water.

As regards the (re)treatment of fluoride salts, processes exist but there are fewer returns than for chlorides, and it seems, for the moment, less simple to set up because it requires the implementation of more and different techniques.

Essentially, five salts have been studied in the United States [1]:

(1) LiF-BeF_2 (FliBe)
(2) LiF-NaF-KF (FLiNaK)
(3) KCl-$MgCl_2$
(4) $NaNO_3$-$NaNO_2$-KNO_3 (HITEC)
(5) LiCl-KCl (IFR, Argonne Laboratory)

It should be noted that some salts are mainly studied in the context of very-high-temperature reactors (VHTRs) and do not take into account the problems of reprocessing. It should also be noted that HITEC is a nitrate with fairly unstable behavior in temperature, so it can be ruled out immediately.

If we take stock of the fluoride salts proposed in 2018 in the various existing MSR projects around the world, we find:

(1) Several projects with FLiBe, including the Chinese project (except on their test loops because of toxicity!) and especially for the nonfast versions (copy of the MSRE). FLiNaK is being considered for the future.
(2) FliNaK (including the Russian project Mosart).
(3) LiF (including the Indian project and Trans Atomic).
(4) NaF/BeF_2 or NaF-KF-ZrF_4 type mixtures.

American startups utilize fluorides when they are in the thermal spectrum, on the basis of the MSRE feedback, and chlorides when they are in the fast burner version (TerraPower and Elysium).

Caution: One must be wary of certain choices often linked to reactor options without salt reprocessing (considered as final waste after a few years of operation without reprocessing).

In conclusion:

- For fluorides, we will look at pure LiF, FliBe, and FLiNaK, which are the main known and studied options.
- For chlorides, we will mainly look at NaCl, KCl, and LiCl, alone or combined.

Operating temperature and U/Pu solubility for fluorides.

This is an important choice parameter, because we try to stay below 700°C for the operating temperatures of the materials.

(1) Pure LiF as solvent

The addition of U and Pu in sufficient quantity to be critical and isogenerating will lead to the formation of a eutectic whose melting temperature will be lower than the melting temperature of pure LiF (see ternary diagram in Fig. 1) [1].

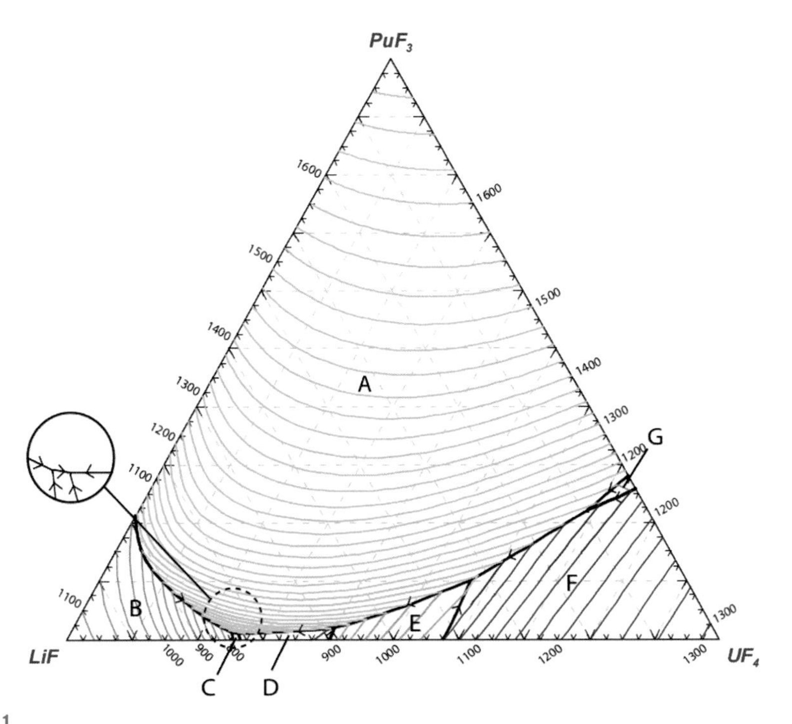

FIG. 1

LiF-PuF_3-UF_4 diagram.

The minimum melting temperature of the LiF-UF_4-PuF_3 salt is 757 K (484°C, orange point on the diagram), for a composition of:

1. LiF: 73.3%
2. UF_4: 25.7%
3. PuF_3: 1%

Initial calculations show that it is necessary to reach a proportion of 6% Pu to obtain criticality and an isogenerating state [1]. The red line in the Fig. 1 corresponds to this proportion of Pu necessary for criticality.

The minimum melting temperature at which the proportion of 6% Pu is maintained and the proportion of heavy nuclei is minimized is represented as the brown point on the diagram.

This operating point corresponds to a molar composition of approximately:

1. LiF: 80%
2. UF_4: 14%
3. PuF_3: 6%

The melting point of this salt is given by the isothermal line highlighted in red in Fig. 1. The temperature of this salt is about 900 K (627°C).

This melting temperature corresponds to the minimum temperature (cold point) that can be reached before precipitation and without operating margin in the face of a cold shock. Thus, the operating temperatures will be higher than the desired 700°C.

Note that it is possible to use the LiF-UF_3-PuF_3 salt, but in this case the lowest melting temperature is at 645°C with 36.4% U and 0% Pu.

(2) FliBe as solvent

If we add BeF_2 to pure LiF, we obtain a salt known as FLiBe and the melting temperature drops significantly, which was the advantage sought in the MSRE with the available materials.

In our case, French legislation would now make any experimental trial, or even any use of beryllium, very difficult because this compound is extremely toxic.

Moreover, the FLiBe is neutronically less interesting for a fast version because it is a neutron moderator, that is, the neutron spectrum is slower.

Finally, the solubility of Pu is limited and reaches less than 1%. Remember that the MSRE only worked with ^{235}U and ^{233}U (it did not work according to the U/Pu cycle) and that this Pu solubility criterion was not important.

This salt is therefore not suitable for the type of reactor that interests us.

(3) FLiNaK

The melting point of pure salt is 454°C. However, the addition of U or Pu in sufficient quantity leads to a notable rise in temperatures.

Initial neutron calculations show that, depending on the quality and quantity of Pu used (use of Pu from PWR or MOX fuel), this molar concentration can vary from 1.5 mol% to more than 3 mol% to reach criticality, which greatly increases the operating temperature of the reactor.

In fact, the solubility of Pu alone is high (up to 15%) and at acceptable temperatures. The same applies to uranium alone. On the other hand, the simultaneous addition of uranium, in the form of UF_4, and of plutonium, in the form of PuF_3, leads to a lowering of the solubilities below 625°C, and in fact notably increases the operating temperature (625°C) to achieve a minimum solubility of 6%. This leads to a T_{max} of more than 700°C at the core outlet if safety margins are taken to prevent a "cold shock" type incident [2].

In FLiNaK, the potential risk of PuF_3 precipitation is linked to the steep solubility slope between 600°C and 700°C. With regard to safety (and therefore the margins necessary to prevent a "cold shock" type incident), it would be necessary to guarantee that the temperature of the salt will never drop below 650°C... which completely annihilates the interest to use FLiNaK with UF_4.

In conclusion, on the solubility in fluoride salts:

On the sole criterion of operating temperature, and the solubilization temperature of U and Pu, it can be seen that:

(1) Pure LiF leads to very high operating temperatures compared with the materials currently available; its use is therefore subject to R&D on the materials.

(2) FLiBe cannot be used because Pu solubility is too low (regardless of its toxicity and the very restrictive legislation on its use).

(3) FLiNaK makes it possible to reach more suitable temperatures with correct solubilization rates for a single species: U or Pu. But with U/Pu coupling, there are risks of precipitation and therefore of uncontrolled deposits of Pu between the operating temperature and the melting temperature. On the other hand, in the "Pu burner" version alone, this solubility seems to be acquired in the required temperature range.

Treatment of fluoride salts

(1) Existing feedback experience

At the level of (re)processing feedback, in the metallurgical industry of processing rare earth metals into fluoride, there is only LiF, all other fluorides being to be avoided owing to the methodology used and vis-à-vis the elements that we want to extract.

Currently, for fluorides, we have no reliable data on transuranics, no demonstration of feasibility on reconstituted material, no associated technology, and no verification on irradiated combustible salt.

It should be recalled that the FliBe from the MSRE has never been reprocessed (the uranium has just been extracted by fluorination).

(2) Electrolysis/reminders on redox potentials

When looking at the oxidation/reduction potentials of the different species, the order of reduction is uranium/neptunium/plutonium/americium/curium/lanthanides.

In fluoride, at 800°C, the reduction potentials of K and Na (and also Be) do not allow the lanthanides to be extracted in FLiNaK—K and Na are reduced beforehand, and therefore the solvent salt is destroyed. We see in the table below that BeF_2 does not necessarily present a better option.

The conclusion to be drawn is that we can extract U and Pu by electrolysis or by reductive extraction, but then if we continue we will "unwind" the salt by first extracting K and Na (or Be), which will force the temperature to rise to about 900°C with almost pure LiF (raising the question of what materials are suitable for treatment at such a temperature) and which would lead to losing the salt.

(3) Fluoridation

Fluorination would also make it possible to extract U, part of the Pu, and Np. On the other hand, the experience feedback on the fluorination of Pu is not favorable. Compared with electrolysis, it is a "gas factory" with deposits everywhere and criticalities that are difficult to control in the vapor/condensation phases.

Studies carried out in the 1960s and early 1970s at Oak Ridge National Laboratory (ORNL) [3] showed that, while the recovery of uranium could be quantitative, that of Pu is much more difficult, as large excesses of fluorine are necessary to oxidize PuF_4 to PuF_6 (69 mol of F_2 per mole of Pu).

Only a method consisting of dropping microdrops of salt (100 μm in diameter) into a column flushed with fluorine at 640°C made it possible to obtain good results (<90% of the Pu recovered over 1.5 m of column; calculations recommend a column of at least 3.5 m).
The differences in volatility between U and Pu would require recovering first U, then Pu (Ref. [3]: ORNL DWG 68-888 and ORNL 4224).
So, electrolysis seems to be a simpler way to implement industrially, in particular for the extraction of Pu.

(4) Selective precipitation by injection of oxides
The injection of oxides could make it possible to extract it first (Note: it can also be extracted in metal form by cathodic reduction, or even by adding metallic uranium).
Precipitation brings into play other equilibria, and the oxides and/or oxyfluorides of lanthanides are easier to form than the oxides of Na, K, and Li. This is why they may be separated by precipitation. However, it is not always simple because oxyfluorides can be quite stable and soluble.
Regarding Am and Cm (which we want to keep in the combustible salt for incineration), the problem is that they will also precipitate when we precipitate the lanthanides. One solution would be to precipitate Am and Cm before the lanthanides. Because the goal is to extract lanthanides in the absence of actinides, a proportion, even a large one (50%), of the lanthanides following the actinides is not a problem since these elements return to the reactor.
Selective precipitation deserves to be tested. It is a matter of adding oxidants little by little to precipitate only part of the elements. As this precipitation would take place according to the respective affinities of the elements, it would be selective.
Finally, the possibility of extracting the actinides curium and americium remains to be demonstrated. If we accept not extracting them, we can imagine proposing a process extracting only U and Pu, but with negative effects on the radiotoxicity of the final waste. In this case, it would be possible to consider a second aqueous step to recover the minor actinides, but this assumes complete dissolution in the aqueous phase.

(5) Distillation
Distillation is a possible way of extracting salt (see examples in China).
Two methods are possible. Distillation can be carried out first, followed by treating the remaining residues by conventional hydroprocesses (GANEX).
Or we can extract the U and the Pu by other processes, then carry out a distillation.
There are then two possible outcomes:

- What remains is waste.
- An additional extraction is carried out to recover the minor actinides (AMs); this complicates the process but makes it possible to reduce the radiotoxicity of the final waste and to have a more favorable strategy for developing MSRs.

Conclusion on the treatment of fluoride salts:

Today, the extraction of U and Pu by fluorination is complex to implement industrially and still not operational for Pu. Industrially, electrowinning would be preferred. However, it is necessary extract Na and K before "unwinding." However, this prevents the extraction of AMs and lanthanides.

Selective oxide precipitation could be effective in extracting zirconium, or even an Ln/AM mixture.

We would then arrive broadly at the following process, all the steps of which remain to be validated:

Step 1: The Zr (or part of it) is removed by precipitation with oxides (or even by cathodic reduction or addition of uranium metal). The Zr goes to the final waste.
Step 2: We leave U/Np/Pu preferably by electrodeposition, because we do not add any pollutant to the salt that is reprocessed.
Step 3: These three elements are returned to the core by adjusting their concentrations if necessary.
Step 4: A gradual precipitation is carried out in an attempt to remove the AMs from what remains of the salt being treated. These extracted AMs are returned to the core.
Step 5: After distillation, the salt is recovered and returned to the core.
Step 6: All the residue remaining after distillation is sent to final waste.

The entire proposed process remains very complex and requires extensive validation work.

Operating and U/Pu solubility temperatures for chlorides

Many salts can be imagined, including NaCl, LiCl, KCl, CaCl, CsCl, and $MgCl_2$.

By way of example, the melting temperatures are as follows for three mixtures:

- NaCl-LiCl with the proportions 30–70 mol% and Tf = 560°C.
- $CaCl_2$-NaCl with the proportions 50–50 mol% and Tf = 500°C.
- $CaCl_2$-LiCl with the proportions 40–60 mol% and Tf = 500°C.

As LiCl/KCl with the proportions 59–41 mol% has an interesting melting point of 352°C, it has been studied more than the others.

LiCl-NaCl-KCl, with a eutectic temperature of 317°C, also has an interesting melting point.

Refs. [4–6], and show the general potential of chlorides when working at lower temperatures, around 500°C, and the good solubility of U/Pu coupling. Indeed, the solubility of Pu, even in the presence of uranium, remains higher than the values necessary for criticality throughout the range of operating temperatures.

For dissolution in the LiCl/KCl eutectic (lines at 58 mol% LiCl), the solubilities at 500°C are 7.6 mol% for $PuCl_3$ and 39.4 mol% for UCl_3. For a lower LiCl concentration (around 40% mol%), it is possible to dissolve more than 40 mol% of $PuCl_3$ [7].

Apart from the often-studied LiCl/KCl, it should be noted that NaCl alone has interesting characteristics. The creation of a eutectic with Pu and uranium makes it possible to significantly lower the operating temperature.

Neutron analysis of the different types of salts

Neutron simulations with different types of salts were carried out in the U/Pu cycle, with depleted uranium containing 0.2% ^{235}U and plutonium from the reprocessing of irradiated uranium oxide.

Regarding NaCl, the ternary diagram in Fig. 2 shows that melting temperatures slightly above 500°C can be reached with an NaCl, $PuCl_3$, and UCl_3 eutectic. Under these conditions, and with the calculation assumptions used, criticality is reached with 8.06% Pu if the core volume is 9 m^3 and 6.71% if the core volume is 18 m^3.

^{37}Cl has a very low neutron absorption rate. It should also be noted that this element has a large ionic radius, which implies that at a fixed volume there are far fewer atoms in a chloride than in a fluoride. To achieve regeneration, it is necessary to simultaneously increase the volume of the core and the proportion of actinides in a chloride fuel salt. This characteristic could prove to be advantageous since it will allow, for a given mass of fissile, to use a higher volume of combustible salt to evacuate the heat.

The production of ^{36}Cl is limited to 1.2 mol/year, which is probably acceptable and, of course, assuming an initial enrichment of 100% in ^{37}Cl.

Calculations made with chlorides of potassium, cesium, and calcium show prohibitive capture rates compared with sodium, which a priori excludes the use of these salts as they are unfavorable to regeneration.

Finally, it appears that the best salt, from a neutronic point of view, is NaCl.

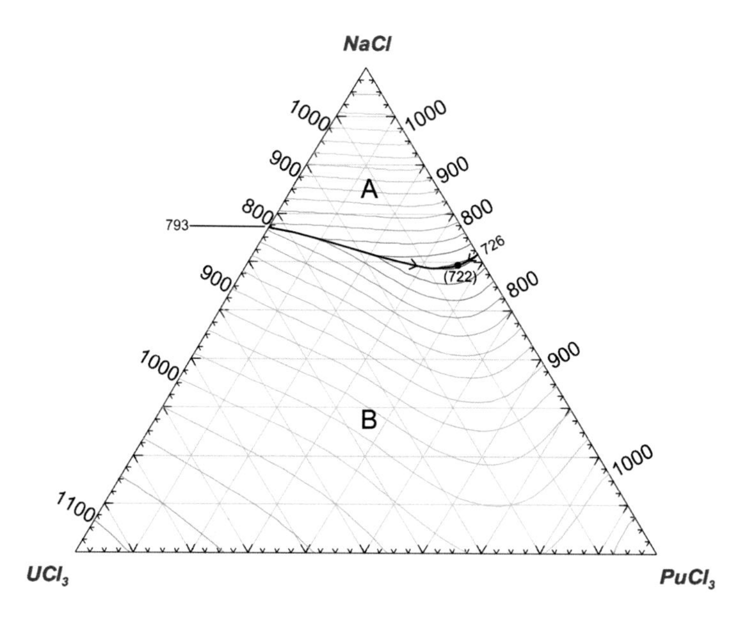

FIG. 2

Melting point of the ternary mixture NaCl-UCl_3-$PuCl_3$ [5]. The ends of the *black line* correspond to *T*=520°C/32 mol% UCl_3 (left) and *T*=449°C/36 mol% $PuCl_3$ (right).

Treatment of chloride salts

For chloride, the feedback from (re)processing is significant. For example, rare earth metals can be obtained in different chlorides (NaCl, KCl, LiCl, etc.); this is the case for mishmetal, for example. In addition, there is industrial experience for reprocessing irradiated metallic fuels and is available for chloride at the Idaho National Laboratory. Also, the Russian process reprocessed tons of irradiated oxide fuels into chloride at the Research Institute of Atomic Reactors. This feedback is a very important asset to be exploited in the development of an MSR using a chloride salt. Moreover, at the level of (re)treatment, chlorides have the very important advantage of being able to use selective electrolysis techniques to be able to extract the desired products in the desired order: Zr, U, Pu, minor actinides, and lanthanides, without "unwinding" the fuel salt. The method of reprocessing chlorides by pyrochemistry therefore seems simpler, more proven, and with available operational industrial techniques, compared with the reprocessing of fluorides.

However, the most important advantage of chlorides is the possibility of solubilization in water, which will allow the use of existing techniques and reprocessing facilities, of La Hague, for the extraction of U, Pu, and minor actinides before their reinjection into the core (remember in this respect that fluorides are prohibited at La Hague owing to corrosion of the materials in the current reprocessing chain).

In conclusion, chlorides offer interesting possibilities by pyrochemistry, but above all seem to allow the use of techniques in an aqueous medium currently available in France.

Disadvantages/advantages of chlorides.

Chlorides have a number of characteristics that are not necessarily prohibitive and that may prove to be advantages:

(1) The physical characteristics (conductivity, exchange coefficient, etc.) are less favorable for chlorides than for fluorides. At equal power, the overall volume of fuel salt required will therefore be greater.

(2) They have a harsher neutron spectrum than fluoride salts, which has two consequences: radiation damage is higher on structures than with fluoride salts. However, this drawback can be solved by using protections on or in front of the structures concerned. In addition, a faster, harder spectrum allows for a more efficient MSR burner (it is for this reason that chlorides have been retained in the "MSR burner" options of American startups).

(3) They are soluble in water. This can be annoying in the event of a leak (accident) because there is a risk of leaching with the humidity of the air at room temperature. On the other hand, this solubility with water allows components and structures to be washed, which is certainly an advantage in terms of component maintenance and, therefore, reactor operation.

(4) Production of chlorine-36 from chlorine-35. An initial enrichment in chlorine-37 is necessary to minimize this production. This issue must be addressed for the solvent (e.g., NaCl) but also for solutes ($PuCl_3$, UCl_3, etc.). Furthermore, chlorine-35 is a neutron poison, which also justifies its prior elimination.

(5) Inevitable sulfur production from chlorine-35 [8] (by neutron capture) can lead to risks of corrosion that are currently poorly understood. In general, feedback on the corrosion of materials in chloride reactor operation does not exist. Here too, an enrichment in chlorine-37 would make it possible to limit this production. Note also the strong affinity of sulfur with metals, in particular Ag and Fe, which could induce trapping. It should be noted, however, that NaCl-KCl loops operate inactively (therefore without sulfur production) with 316 L materials [4], without corrosion. In general, it is the impurities produced that can lead to corrosion, and a redox control method will have to be developed to avoid it.

General conclusion and recommendations

For a U/Pu cycle, the use of fluorides does not seem possible as it stands. Indeed, the risks of precipitation of a Pu fluoride in a temperature range between criticality temperature and melting temperature seem prohibitive in terms of intrinsic safety. In addition, the operating temperatures are quite high (which poses a problem for the materials), and the development of a reprocessing scheme seems more difficult and has a fairly low feedback experience.

For a Pu burner, the use of FLiNaK would seem possible with high but acceptable temperatures. However, the corresponding reprocessing scheme would remain complex to set up and validate.

The chloride salts make it possible to operate in a U/Pu cycle with acceptable operating temperatures with respect to the available materials and with sufficient solubilization rates of the U/Pu coupling in the entire temperature range between melting temperature and operating temperature.

In addition, the (re)treatment of chlorides, even if it remains to be clarified, seems simpler and more industrially proven, in particular by methods in an aqueous medium such as those available today at La Hague.

Chlorides therefore seem to be better candidates for the context that interests us (fast spectrum for a U/Pu cycle, with multirecycling and subsequent possibility of incineration of minor actinides).

It should also be noted that the two American startups working on "rapid" burners use chloride salts.

Several chloride salts remain theoretically possible, but neutron simulations show that NaCl is the best candidate for the main component.

NaCl with U and Pu forms a eutectic allowing an acceptable operating temperature and well below 800°C of its melting temperature. On the other hand, this imposes certain concentrations of U or Pu; otherwise the temperatures will rise. In addition, if the reprocessing begins with extraction of uranium, the temperatures will rise to around 800°C.

The addition of a salt making it possible to create with NaCl a eutectic with an intrinsically lower melting temperature is an option to be considered. This salt must be chosen so as not to penalize too much at the neutron level. Among the possible salts, $MgCl_2$ appears as a good candidate. In addition, its greed for water brings an interesting element of intrinsic safety, to avoid selective deposition of plutonium in the event of incidental moisture entry.

At this stage, NaCl remains the reference solvent for carrying out subsequent design studies of a fast MSR in the U/Pu cycle. Prior isotopic separation of Cl is necessary. The addition of another salt (such as $MgCl_2$) to create a lower melting temperature eutectic is an interesting option.

References

[1] ICAPP, Molten Salt Reactor to Close the Fuel Cycle: Example of MSFR Multi-Recycling Applications, 2019.

[2] E. Capelli, O. Beneš, R.J.M.K. Onings, Thermodynamic assessment of the LiF–ThF_4–PuF_3–UF_4 system, J. Nucl. Mater. 462 (2015) 43–53.

[3] A. Degtyarev, A. Myasnikov, L. Ponomarev, Molten salt fast reactor with U/Pu fuel cycle, Prog. Nucl. Energy 82 (2015) 33–36.

[4] A. Lizin, S.V. Tomilin, V.S. Naumov, V. Ignatiev, N.Y. Nezgovorov, A.Y. Baranov, Joint solubility of PuF_3 and UF_4 in a melt of lithium, sodium, and potassium fluorides, Radiochemistry 57 (2015) 498–503.

[5]" Thermodynamic Evaluation of the NaCl–$MgCl_2$–UCl_3–$PuCl_3$ System " O. Benesˇ, R.J.M. Konings, n.d. European Commission, Joint Research Centre, Institute for Transuranium Elements, Karlsruhe, Germany.

[6] S. Ghosh, B.P. Reddy, K. Nagarajan, K.C.H. Kumar, Experience Investigations and Thermodynamic Modeling of KCl–LiCl–UCl_3 System, Elsevier, 2014.

[7] W. Zhou, Integrated Model Development for Safeguarding Pyroprocessing Facility (thesis), Ohio State University, 2017.

[8] M. Taube, n.d. Fast Reactors Using Molten Chloride Salts as Fuel (§4.7). EIR Report No. 332.

[9] ORNL reports n.d. "ORNL DWG 68-888 and ORNL 4224".

Index

Note: Page numbers followed by *f* indicate figures and *t* indicate tables.